R.D. Toosey.

PROFITABLE FODDER CROPPING

PROFITABLE FODDER CROPPING

R. D. TOOSEY
MSc (Agric), NDA

FARMING PRESS LIMITED
FENTON HOUSE, WHARFEDALE ROAD, IPSWICH, SUFFOLK

First Published 1972

ISBN 0 85236 026 6

This book is set in 10pt. on 11pt. Times and is printed in Great Britain on S.E.B. Antique Wove paper by The Leagrave Press Ltd., Luton and London.

CONTENTS

ILLUSTRATIONS

FOREWORD

by SIR EMRYS JONES, BSc (Agric)

Director General, Agricultural Development and Advisory Service, Ministry of Agriculture, Fisheries and Food

FOR MANY years now our traditional root and fodder crops for stock have tended to be overshadowed and even neglected in the intensive drive towards better grass and better grassland management. We are, however, entering upon an entirely new situation.

I am sure that livestock producers would be wise to seek a wider variety of fodder and better conservation of quality winter feeds. Many arable farmers will certainly be looking for new cleaning or break crops. All will be searching for increasing productivity and profitability of their farming systems.

The new varieties of fodder crops available, the new methods of production and the better ways of using them can help all farmers to achieve their goals. I am equally convinced that high quality fodder crops of all kinds will in the future make a massive contribution to an efficient agriculture in this country.

This book comes at a most opportune time. It draws together the most up-to-date information available and suggests many ways in which suitable fodder crops can help farmers to meet the challenge of the years ahead.

EMRYS JONES

October, 1972

PREFACE

The situation today is not dissimilar to that which prevailed during the Agricultural and Industrial Revolutions of the last century. We have a rapidly expanding urban population which must be fed economically and without undue reliance on imported foods. The high productivity of the arable fodder crops allowed them to play a most valuable part in arable and livestock production in those early days; their importance declined with changing farming systems and high labour costs.

New methods of production, new varieties and other recent advances in technology could enable arable fodder crops to make a really valuable contribution to present day agriculture. Given skilled management, the exploitation of the full potential of these crops could bring about an increase in livestock production far outstripping anything we have seen before and this without sacrificing cereal output.

The purpose of this book is to bring together some of the information now available on the production and utilisation of the arable fodder crops; also to suggest to farmers some ways in which they may increase the productivity and profitability of their farms by incorporating suitable fodder crops into their individual farming systems.

July, 1972

R. D. Toosey

ACKNOWLEDGEMENTS

I WISH to express my gratitude and thanks to the many colleagues, friends and organisations who have helped in the preparation of this book by contributing information, by supplying material and photographs and by constructive criticism of the text.

I especially wish to thank my colleagues at Seale-Hayne Agricultural College: Mr S. L. Hewitt, Clerical Technician, for taking and printing the photographs for Plates 2-11 inclusive and Plate 13 and for his technical assistance in reproducing tables and figures; Mr K. C. Vear, Head of the Botany Department, for reading and criticising Chapters 1, 2, 3, 5, 6 and 9, and for his invaluable help and encouragement; Mr J. L. Jemmett, Regional Trials Officer, NIAB, Seale-Hayne, for reading and criticising Chapter 6, for his generous technical help and for supplying materials for some of the plates; Mr S. E. Niemeyer, Lecturer in Farm Management for reading and criticising Chapter 10, and Mr D. C. Henderson, Senior Lecturer in Animal Health, Mr D. J. Iley, Lecturer in Zoology, and Mr E. F. Thorpe, Senior Lecturer in Animal Husbandry, for their help with appropriate sections in Chapter 4. I also wish to thank especially Mr K. P. Moase of Traceys Longcombe, Totnes, for reading and criticising Chapters 7 and 8.

I am indebted to Dr P. S. Wellington, Director of the National Institute of Agricultural Botany and his colleagues at headquarters, Mr L. A. Willey, Officer for Fodder and Root Crop Trials and Mr D. T. A. Aldrich, Officer for Grass and Legume Trials for their very considerable help and for their kindness in supplying information on fodder crops and grasses; also to Mr A. D. Rivett, Regional Trials Officer, NIAB, Cockle Park, Northumberland, for his help with the tables in Appendix I. I am also grateful to the NIAB for permission to reproduce their copyright material.

I also wish to express my appreciation to Dr H. P. Allen, Mr M. G. Denning and their colleagues at Plant Protection Limited for reading and criticising Chapters 7 and 8 and for their help and advice on direct drilling so freely given; thanks are also due to Plant Protection Limited for supplying the photographs of direct drills and direct drilled crops for use on the dust jacket and Plates 12 and 14.

The help of the following members of the Agricultural Development and Advisory Service is greatly appreciated: Mr P. J. Attwood, WRO Liaison Officer for reading and criticising Chapter 8, and Messrs D. N. Beasley; C. W. C. Dawkins; S. C. Melville; R. Muscutt; E. I. Prytherch; J. Rhodes; G. Sharp; R. Thomas; G. A. Toulson and J. G. Wilson for their technical help.

I would also like to thank the Maize Development Association, for allowing the reproduction of material in Tables 22 and 23 and Figure 3; Mr W. F. Raymond, formerly of the Grassland Research Institute, Hurley, for supplying material and technical data on maize and whole crop cereal silage reproduced in Figures 4 and 5; Mr J. Reid and Mr R. W. T. Hunt of the West of Scotland College of Agriculture, for allowing publication of material in Tables 32 and 33; Mr D. Morrison of the North of Scotland College of Agriculture for supplying information on swedes, and the Ministry of Agriculture, Fisheries and Food, Exeter, and the Department of Agriculture and Fisheries for Scotland, for supplying data on crop acreages.

July, 1972   R.D.T.

Chapter 1

INTRODUCTION

FODDER CROPS—root and green forage crops—have long been appreciated as an excellent source of succulent food for fattening and milk-producing stock during the autumn, winter and early spring months when grass is largely dormant and unproductive.

Traditionally these crops, especially the roots, have been of great importance in both mixed and predominantly arable systems of farming. However, in spite of the increase in the number of beef and dairy cows in England and Wales between 1950 and 1970 and in the number of breeding sheep up to 1966 (Table 1), the area of tillage land devoted to fodder crops declined dramatically over the same period.

Thus by 1970 the area of fodder crops recorded at the annual June 4th census by the Ministry of Agriculture, was a little over one-third of that grown in 1950 and occupied a mere 3·4 per cent of the total tillage area (Table 2). Although the rate and timing of the decline differs, a comparable picture emerges in Scotland.

Such a drastic reduction in the acreage of fodder crops is probably due to a combination of factors, which must include sweeping changes in farming systems, the frequently heavy manual labour involved in production, especially of roots, problems in controlling

TABLE 1. Numbers of Breeding Cattle and Sheep (000) in England and Wales

	1950	1960	1966	1970
Total Number of Dairy Cows*	not returned separately	2595	2631	2714
Total Number of Beef Cows*		518	596	667
Total Number of Breeding Cattle	2951	3113	3227	3381
Total Number of Breeding Ewes†	5856	8798	9643	8394

* Excluding heifers in calf with first calf.
† Including shearling ewes to be put to the ram.

June 4th Agricultural Census

15

weeds, acute difficulties of utilisation and the great advances made in the growing and mechanical handling of the grass crop in recent years.

TABLE 2. Area of fodder crops (000 acres) in England and Wales

Crop	1922	1950	1960	1970
Turnips and swedes	821*	300	205†	106†
Mangolds	456	267	127	24
Kale	} 137	} 233	355	142
Cabbage savoys and Kohl rabi			19	8
Rape or cole		132	79	46
Maize other than for grain	—	—	—	2
All other fodder crops excluding mustard lucerne and grass	135	44	19	24
TOTAL FODDER CROPS	1550	976	804	352
TOTAL TILLAGE	8957	10390	9257	10367
TOTAL FODDER CROPS AS PERCENTAGE OF TOTAL TILLAGE	17·3	9·4	8·7	3·4

* Includes turnips and swedes for human consumption.
† Includes fodder beet.

June 4th Agricultural Census

FARMERS' LACK OF INTEREST

When viewed in the context of a rapidly shrinking labour force and rising wages, the swing away from the fodder crops is understandable—at least on the basis of traditional methods of production. To these reasons may be added the almost entire lack of interest in their potential by many farmers, most of their advisers and numerous agricultural merchants, where root and brassica seeds formed an insignificant part of their trade. Similarly, only recently have the scientific plant breeder, the agricultural engineer and the manufacturer of herbicides started to make their contributions in the production of new varieties and new labour-saving methods of production, utilisation and weed control.

In contrast, much attention has been focused on the grass crop and large sums of money have been spent by government and industry on research and development.

Progress has been dramatic. Many improvements such as leafier and more persistent varieties of grass, greatly increased fertiliser usage, especially of nitrogen, the reseeding of poor pastures and improved methods of grazing control have combined to raise the productivity of grassland out of all recognition and extend the length of the grazing season substantially.

However, the major factor in the field of grassland management which helped to bring about the decline in the acreage of fodder crops was probably the widespread adoption of the forage harvester. This machine, combined with the many other advances in fodder conservation and utilisation, such as self-feed silage, has made it possible for farmers to rely on grass as their sole source of winter fodder; whether the results have been entirely satisfactory is quite another matter.

Fodder crops still have an extremely useful part to play in British agriculture, which may be best appreciated by an examination of the weaknesses of grassland and intensive cereal farming.

LIMITATIONS OF GRASS CROP

It cannot be reasonably disputed that intensively managed grassland normally offers the best and cheapest source of grazing during the spring and much of the summer period. Short, young spring grass is highly digestible, rich in protein and is a 'production' food par excellence which will fatten sheep and cattle and will support heavy milk yields without supplementation by concentrates.

Unfortunately the feeding value of grass tends to fall off as the summer proceeds and supplementation with other food becomes increasingly necessary to maintain full production. With expert management a high level of milk production can be maintained on grass alone well into the summer but, frequently, there is a tendency to over-estimate the feeding value of the grass, particularly in the late summer months.

Late autumn grass, especially from young leys, tends to have a relatively low feeding value and is generally inadequate for fattening stock unless it is supplemented; similarly it does not provide much more than maintenance for dairy cows. In a wet summer grass can reach this stage by mid-August.

Dry weather may also present severe problems from July onwards, especially on the lighter soils throughout the south of England. A few acres of early sown kale can then be most valuable to dairy cows.

Similarly, some farmers with late lambing flocks change to rape in the early autumn; this practice also provides a move to ground which is free from stomach worms and other internal parasites. A combination of Italian ryegrass, grazed winter cereals and brassica fodder crops is particularly valuable in the latter respect, as they

B

provide a continuous change to clean land and permit very high stocking rates to be achieved with complete safety. On long leys and permanent grass there is always a danger of building up a very heavy worm burden with intensive stocking by sheep for a protracted period.

WINTER GRASS

The productivity of grassland during the winter period, which effectively lasts from five to seven months according to district, is generally low. The grazing season may certainly be lengthened by the application of nitrogenous fertilisers to grassland, but it is not uncommon for the land to be too wet to utilise the grass without damaging the sward unduly. In fact, good management usually necessitates that grassland is only lightly stocked or completely unstocked throughout most of the winter to avoid poaching and over-grazing.

Poached pastures take much of the following summer to recover, while overgrazing in the late winter and early spring seriously delays spring growth and results in reduced yields of grass. In these circumstances a supply of suitable fodder crops keeps stock off the pastures and simultaneously improves the current output of the stock and the future output of the grassland, as well as increasing the stocking rate of the whole farm. Nothing is more ruinous to pasture productivity than a heavy head of outwintered stock, especially when they are fed from tractors and trailers.

The alternative is to provide adequate buildings or yards for over-wintering. This course is essential on wet land and is usually necessary with very productive stock such as high-yielding dairy cows. However, with other types of stock, only inexpensive buildings can be considered. If the floors and walls cannot be adequately disinfected, such yards and buildings constitute a considerable disease risk, especially with lambing ewes.

On the lighter well-drained soils a combination of Italian ryegrass, green fodder crops, cereal stubbles or a pasture ready for the plough makes a cheap and effective method of overwintering sheep, beef cows and dry cattle with minimal capital expenditure.

CONSERVED FODDER

The major weakness of grass becomes only too apparent when the crop is conserved as hay or silage. The resultant product is frequently of very indifferent quality and requires heavy supplementation with concentrates for production—either for milk or fattening. A substantial proportion of the hay that is made has a feeding value no better than that of barley straw, while much of the silage can only provide maintenance or maintenance plus half a gallon of milk per day; both are then of relatively low or very low digestibility. With really skilled management and sophisticated equipment the product may be improved greatly but then only the more productive stock will stand the cost.

It has been shown that, with heavy applications of fertiliser nitrogen, about 250–300 units N per annum, grass is capable of producing high yields of conserved fodder, often in the region of five tons of dry matter per acre per annum; more has been recorded. To do this it is generally necessary to take three cuts in the season, but it is not uncommon for the third cut to be missed.

FIRST-CUT SILAGE

A particular difficulty with the first cut of grass for silage, which is also the heaviest, is the very short time in which the grass gives a good yield combined with a reasonable dry matter content and a high digestibility. If cut too early, the digestibility is high but the yield and moisture content are low and a wet poorly fermented silage may result. Once past the optimum stage—at the start of the emergence of the flowering head from its sheath—the digestibility falls rapidly. Silage cut when a high proportion of the heads are well emerged and the crop is deficient in leaf, which includes many of the typical grass silages made today, is of rather low digestibility.

In most seasons grass suitable for first cut silage is only available for a period of some 12 to 14 days, so that the usual difficulty is the sheer physical task of making a large enough weight of silage in the short period that is available. The job still presents considerable problems when the farm is highly mechanised, especially in rainy weather and when breakdowns occur.

Furthermore, wilting is usually necessary in order to obtain a silage with an adequate dry matter content. This practice is frequently ineffective in the higher rainfall areas and in a wet May, as some owners of tower silos in the west of England have found to their cost.

Fodder crops can help to ease the grass conservation problem either by reducing the need for conserved fodder or by providing a suitable alternative raw material for conservation. Thus roots and brassica green fodder crops, which retain a high digestibility over a long period can replace a substantial part of the conserved winter ration, while maize and sometimes whole crop barley and wheat may be used as an alternative to grass silage.

Maize shows promise in suitable areas, giving a product with a digestibility equal to the best grass silage and five tons or more of dry matter per acre in only one cut. As maize silage is made in late September or early October, it does not clash with the making of grass silage. Neither crop requires wilting before ensiling and they are easily handled mechanically.

FEEDING SHEEP

The feeding of sheep poses a special problem, as they must have a palatable food with a high digestibility and a low fibre content. Roots and succulent green forage crops fulfil these requirements exactly.

Apart from the question of cost, considerable problems arise when large quantities of conserved fodder are fed. Hay must be of high quality. Silage is an excellent food for ewes but the grass must be cut and ensiled when short and very leafy; the problem is usually to ensile enough of this type of material. 'Cow silage' is quite unsuitable for sheep.

With the heavy labour of carting silage and the frequent mess around the feeding racks, there is everything to be said for relying mainly on cheaply grown succulent root or fodder crops when fresh grass cannot be grazed; these may be supplemented with good hay where necessary.

CEREAL GROWERS' RETURNS

The low financial returns experienced by some intensive cereal growers may often be attributed to reduced yields, typically caused by soil-borne and foliar diseases, grassy weeds or deteriorating soil structure. Low incomes may also be due to insufficiently intensive use of the land, which hardly seems sensible with heavy overheads and high land prices.

The introduction of 'break crops' into the cropping sequence, which means the introduction of some form of crop rotation, can prove to be a very effective answer to the problems of weeds and soil borne diseases. However, in the latter case, the land must be kept free from wheat and barley for two consecutive years if the break is to prove effective.

All too often, 'break crops' grown for direct sale, such as oats, oilseed rape, beans, peas or herbage seed, have either a limited market, may be limited by quota or contract or are unable to offer a cash return comparable to that of a reasonable cereal crop. Maize for grain shows promise but can only be grown in a limited area and more experience must be gained before the crop can be grown with absolute confidence.

ALTERNATIVE LIVESTOCK

The alternative is to introduce livestock into the cropping system and use a ley as the 'break crop'. A grass ley of two to three years' duration included as a regular feature in the rotation acts as an excellent buffer against many of the soil, weed and disease problems —farmers who practice this usually have little trouble. Technically the ley is very efficient; the difficulty is often to utilise the resultant grass profitably.

If dairying can be introduced, intensive stocking of the ley with cows can make the system far more profitable than cereal production alone.

With beef and sheep, traditional stocking rates are not profitable; it is necessary to use high rates of nitrogen, controlled grazing and to obtain a really heavy stocking density. The latter may be further improved by the introduction of catch crops, which are snatched between two cereal crops when the land would otherwise be lying idle. The quick growing brassicas are particularly suitable for the purpose.

Further intensification may be achieved by grazing winter cereals and by the use of Italian ryegrass; in some systems the latter is grown as a 9–15 month ley and may then replace the longer ley.

The inherent beauty of such a cereal-ley-catch crop system is that maximal output and cash returns can be obtained from both crops and livestock at the same time. There is very little 'idle time' between crops and a minimal area is required for the provision of fodder. Crops, grass and livestock help each other; soil fertility and yields may be improved while the livestock is provided with pasturage free from parasites and diseases.

On many farms, fodder is also obtained as a by-product of crops grown primarily for direct sale. Such crops include grass and clover leys for seed production, sugar beet, brussels sprouts and cabbages grown for human consumption. These fit well into any sheep keeping system and vice versa. Leys for seed production may be supplemented by stubble-sown brassicas, while sprouts and cabbage may be augmented with a catch crop of Italian ryegrass. There are numerous interesting combinations.

The current position and the possible future trends of fodder crop production may be best assessed by examining the alterations in the acreage of these crops that have occurred in the past hundred years. The pattern differs greatly from one crop to another.

TURNIPS AND SWEDES

The recorded acreage of turnips and swedes grown in Great Britain reached its peak in 1870, when 2,211,000 acres were grown.

Since then, the area has declined practically without pause until only 247,000 acres were grown in 1970, of which 77 per cent was grown in Scotland, Yorkshire and the English counties further north. The climate in these areas is more suited to growing heavy crops of swedes than the hotter areas to the south.

While the acreage of turnips and swedes has declined massively in most parts of the British Isles, the fall has not been evenly spread. Although still heavy, the reduction in acreage has been least in the traditional swede-and-livestock areas of the North and West; in these areas much of the decline is probably attributable to the heavy labour requirement of hoeing and singling the growing crop and when stored, to the slow laborious job of harvesting and feeding the roots to the stock. In those areas, which are so well suited to growing good crops of turnips and swedes, any advances in the mechanisation of these crops are likely to be met with considerable interest.

Thus, in Brecon and Radnor, where the technique of drilling to a stand has been widely taken up, the acreage is not far short of that grown in 1922 and actually increased from 5,228 to 6,478 acres between 1950 and 1970. For similar reasons the number of turnip harvesting machines in Scotland increased from 2,066 in 1961 to 3,500 in 1971.

In contrast, farmers in the arable areas south east of a line drawn approximately from The Wash to Portland Bill now grow only a few thousand acres of swedes and turnips as maincrops; swedes are, therefore, almost extinct in the south eastern half of England, where once the root break and the arable sheep flock were the central features of the farming system. The massive extent of the decline is evident from the acreages of turnips and swedes grown in 1970 in five counties with substantial arable acreages, which were leading producers of turnips and swedes in 1922:

	Area of turnips and swedes *(000 acres)*	
	1922	1970
Dorset	22·4	0·7
Hampshire	28·5	1·0
Lincs. (Lindsey)	75·3	5·0
Norfolk	84·9	2·0
Nottinghamshire	21·0	0·7

In traditional arable farming, roots, which were predominantly turnips and swedes, constituted the 'fertility restoring' and 'cleaning' break in the rotation. After the 1914-18 war these crops were replaced in many arable areas by crops such as sugar beet which provided a direct cash return and at the same time also gave a good supply of fodder or green manure from the tops.

Concurrently, the replacement of the excreta of the sheep folding the roots and the dung of yard-fed cattle by chemical fertilisers, together with the universal adoption of modern selective herbicides, which turned cereals into cleaning crops for broadleaved weeds, gradually removed the cultural need for the traditional root break in arable systems. Crops were then grown largely on their individual economic merits and the traditional 'root break' was replaced by whatever crop would show a satisfactory financial return.

It is also worth noting that the traditional 'root break' was grown as a maincrop and thus occupied the land for a whole growing season. If maincrops are to prove economically viable, it is essential that they should produce a good return, either in the form of a large bulk of utilisable fodder or direct cash. In the South and East turnips and swedes did neither; the very high risk of powdery mildew precluded early sowing and yields were generally low. It is thus almost certain that the disappearance of swedes and maincrop turnips from the South-East is permanent; other crops give far better yields or are financially much more rewarding.

MANGOLDS

Mangolds never achieved the same importance as that of turnips and swedes, but the decline in acreage did not really gather momentum until after 1945, when the acreage of the crop then fell from 307,000 to 24,000 acres in 1970. The causes were no doubt primarily due to the high labour requirement of growing and handling the crop, allied to the widespread adoption of the forage harvester for grass. However, some growers still produce amazingly heavy yields from this crop, which is remarkably useful for ewes and lambs during periods of hard frost and dry springs. Yields of over 60 tons per acre are by no means impossible under the best conditions. Although mangolds are essentially a maincrop they can be justified in certain circumstances.

CABBAGE, KALE AND RAPE

The difficulty in commenting upon the brassica green forage crops (cabbage, kale and rape) is heightened by the partial or total grouping of these crops in the agricultural returns prior to 1960 and by the increasing tendency for the kales and rapes to be sown as catch crops —a crop snatched between two main crops when the ground would be otherwise unoccupied—rather than as main crops.

As a considerable proportion of the kale and rape grown in England and Wales, appears to be sown after the collection of the returns on June 4th, a significant part of the present-day acreage of these crops is no doubt returned as the previous crop, such as grass, early potatoes or cereals. It is therefore almost certain that the census substantially underestimates the area devoted to quick-growing brassica crops; the extent of the inaccuracy is likely to depend on how late the bulk of each particular crop is sown.

While it appears that a considerable proportion of the kale crop, especially crops direct drilled after grass in the South-West of England, is sown after June 4th, a very high proportion of the rape, especially on arable farms, and probably *all* the stubble turnips are sown late and thus not included.

Comments from some merchants on the amounts of seed sold to farmers certainly seem to suggest that there is a gross under-estimation of the current importance of these crops. It is also worth noting that no distinction is made in the returns between different types of kale.

Cabbages, which once probably occupied the major part of the green crop acreage, are now of minor importance, while kohl rabi may well have been out of cultivation for approximately 50 years.

Kale's Importance

The importance of kale increased steadily up to and after the 1939-45 war, as it was a useful succulent food for sheep and dairy cows.

The thousand head type was used mainly for folding with sheep and was frequently sown in alternating blocks of rows with swedes.

Marrow stem kale was used almost entirely for cows; some growers singled it and it was even transplanted in some instances. Both practices were quite unnecessary and have since been shown to be very undesirable. The crop was mostly cut by hand and carted to the stock.

The adoption of the electric fence allowed marrow stem kale to be grazed in situ and greatly reduced the labour required for utilisation; kale thus became a low-cost crop which was apparently a very attractive proposition. The result was a most striking increase in the kale acreage, which was particularly marked in the South-West, notably in Devon and Cornwall, where the area increased by approximately $2\frac{1}{2}$ times.

The reaction soon set in. The crop was frequently planted on unsuitable fields and the loosening of the soil by ploughing and conventional cultivations made the land particularly susceptible to poaching. The wet autumns of the early 1960s showed up the weaknesses of winter grazing only too clearly. In many fields conditions were quite appalling, the cows often wading hock-deep in a poached morass which became almost impassable around gateways.

Quite apart from the fearsome conditions in the field, life in the cowshed presented a sorry and continuous tale of filthy udders, sore teats and lame feet. Furthermore, as a result of overfeeding kale and the omission of proper mineral supplementation, many herds experienced considerable trouble from metabolic disorders and heavy outbreaks of infertility, especially among the higher-yielding cows.

Rape in Decline

There is little doubt that there has been a heavy decline in the acreage of rape grown in most areas in the last 20 years; probably a considerable area of the crop has been replaced by grass. Rape has also apparently been used less as a pioneer crop and in the reseeding of grassland. Comment beyond this point is difficult, owing to lack of reliable information. However, very substantial quantities of seed still appear to be sold, presumably for catch cropping.

The reduction in the acreage of green forage crops such as vetches, which were once grown extensively to provide spring and early summer keep, may be attributed to the demise of the traditional arable sheep flock and to improved varieties of grass and grassland management. In addition, cereal-legume mixtures are now rarely grown for silage.

However, certain green forage crops are still exceedingly valuable for arable sheep flocks; the modern crops for the purpose include short-term leys or catch crops of Italian ryegrass, cereals grown for grain from which a spring grazing is taken or rye. The first two are either recorded under other headings or go completely unrecorded.

NEW TECHNIQUES

The potential of the arable fodder crops may now be realised by the adoption of techniques which have been developed in recent years. The complete mechanisation of the root crop is now an accomplished fact. In suitable areas drilling to a stand with a precision drill, and the use of herbicides to control annual weeds, is established commercial practice, as in Brecon and Radnor where swedes are an important source of cheap winter fodder for feeding in situ. The mechanical harvesting of the root crop has for some years been normal commercial practice in Scotland, while the ADAS-Walcot self-feed hopper for whole swedes, described in Chapter 4, allows swedes to be self-fed to fattening cattle with minimal labour.

Whether fed in situ or stored, roots are once again a very practicable proposition in areas where they give a worthwhile yield.

DIRECT DRILLING

The introduction of direct drilling, which is dealt with in detail in Chapter 8, has greatly reduced the labour required to grow fodder crops and has made utilisation in situ, a more attractive proposition on all well-drained soils.

The technique is widely practised commercially and 19,500 acres were estimated to have been direct drilled in the South-West of England alone in 1971 (15,800 acres of early sown brassicas and 3,700 acres of brassicas and Italian ryegrass sown in cereal stubbles). The technique is especially well-suited to brassicas and cereals.

In recent years there has also been an increasing awareness of the potential of the long established practice of catch cropping, especially in the south of England.

The kales, rapes and stubble turnips in particular must be eaten when in a young succulent stage; they do not require a long growing season and are thus best grown as catch crops. If left too long these crops merely become woody, unpalatable and much of the crop is wasted. The increased use of catch crops can increase financial returns by making the farm more intensive.

The fodder crops are capable of very high physical output. However, it must be emphasised that they should not be regarded as competitors to the grass crop but as valuable complementary crops. If fodder crops are carefully chosen and built into the farming system, whether primarily grass or predominantly arable, the farmer is likely to benefit considerably.

ASSESSMENT OF YIELD AND FEEDING VALUE

Surprisingly, perhaps, it is more than a little difficult to measure the yield of grass and fodder crops. A simple measurement of fresh or green yield is most unreliable as it takes no account of the water content of the crop and hence gives no indication of its feeding value. For example, 30 tons per acre of a crop containing 10 per cent dry matter (DM) gives exactly the same yield of DM, 3 tons per acre, as 20 tons per acre of a crop containing 15 per cent DM.

The apparently large difference in yield is due simply to the greater weight of water contained in the first crop. The water content of the crop decreases as the crop matures and varies greatly according to crop and variety, the weather, the time of day that the crop is sampled and from one year to another. A comparison on a dry matter basis therefore gives a much more satisfactory indication of yield than the fresh yield only.

Although a comparison of the dry matter contents of different types and varieties is perfectly satisfactory in the case of root crops, which are highly digestible and where the digestibility does not vary very much, it is not so with the grasses and the green fodder crops, where fibre, which becomes increasingly indigestible, is laid down in the stems as the crop ages. In these crops there are considerable differences between types and varieties in the digestibility and hence in the feeding value of the stems as the crops become more mature.

A much better indication of feeding value can be obtained by using the in vitro method of assessing digestibility which has to be undertaken in a laboratory with special apparatus and liquor from a fistulated rumen. Quality may then be expressed as the D value (the percentage of digestible organic matter in the dry matter).

Problems

Although the methods of estimating the likely yields of utilisable food material have been improved greatly in recent years, they are still, at best only a useful guide to the relative potential values of different crops. Not only may the actual yield of a particular crop, such as kale, standing in the field vary greatly from one site to another, but the yield that is actually utilised depends on the stage of growth at which utilisation occurs. There may also be large differences in the feeding value of two apparently identical crops which give the same yield.

Food on offer to stock and food that is actually consumed and consumed profitably, are very different things. When a crop is consumed in situ, the amount utilised will depend not only on the condition and palatability of the crop, but also on the extent to which the crop is soiled, which is again dependant on the state of the soil surface, as influenced by such matters as method of cultivation and weather conditions. The amount consumed is also influenced by the quality and quantity of other foods on offer to the stock.

Chapter 2

THE BRASSICA GREEN FORAGE CROPS

THE BRASSICA green forage crops may be sown to provide an almost continuous supply of green fodder from July until the following April. The group includes kale, forage rape, the swede-like kales, fodder radish and flat poll cabbage; turnips may also be sown alone or in mixture to provide tops for green forage.

THE KALES

The word 'kale' is an agricultural description of plant habit as opposed to one of botanical classification. The leaves are arranged loosely, arising from a stem of variable length which may show considerable thickening in some varieties. Side shoots may also arise from the stem, giving the plant a bushy appearance. The leaves, side shoots and in some varieties the stem, provide the edible portion.

The kales are biennials which flower during the year after sowing. There are both agricultural and horticultural varieties of kale but the latter, which include such types as green curled Scotch kale or borecole and cottagers kale, are lower yielding and are of no significance as fodder crops.

The agricultural kales may be divided into two major groups:

1. The 'true' or *Brassica oleracea* kales which include:

   (a) The marrow stem.

   (b) The hybrid.

   (c) The thousand head kales.

2. The swede-like kales, which belong to the species *Brassica napus* to which the swede and the forage rapes commonly grown in this country belong. These kales are in fact rapes and in this text are included with the rape crop.

Marrow Stem Kale

Apart from the hybrid varieties this type has the highest yield potential. A satisfactory commercial crop will usually produce 20–25 tons of green material ($2\frac{1}{2}$–3 tons dry matter) per acre, although late sowings may give considerably less. Green yields of up to 40 tons (5 tons dry matter) have been obtained but such heavy crops are likely to be difficult to utilise in situ.

The greatest proportion of the crop, from 60–70 per cent of the total yield, occurs in the form of stem, which is relatively long and stout and may attain a thickness of up to four inches at low plant densities.

Green and purple skinned types exist, but the latter is now relatively unimportant. Lateral branches or side shoots are rarely produced. There is little recovery after grazing or cutting.

A cross section of the stem shows a succulent pith, surrounded by a thin ring of conducting tissue, which gradually thickens as the crop matures, eventually becoming very woody. If crops are allowed to reach such a condition before folding, the stems are largely neglected by stock and a very high proportion of the crop is wasted. When fed at the correct stage, the stems are palatable, highly digestible and readily consumed. Marrow stem kale is used mainly for feeding to cattle in situ, particularly to dairy cows.

For autumn feeding

Marrow stem kale is grown mainly for feeding during the autumn; it is not particularly frost hardy and should generally be consumed before Christmas or early January at the latest. The crop is sometimes fed as late as March in the milder parts of south-west England, although this practice is risky and hardier types such as the hybrid variety Maris Kestrel are much safer. If marrow stem kale must be left to stand after Christmas, it should not be sown before mid-June and the level of nitrogen application should be restricted to 60 units per acre. Marrow stem kale may also be grown for grazing in August and early September to replace grass in dry situations.

Marrow stem kale is rarely affected by the common races of clubroot and then only under conditions of heavy infestation and severe soil acidity.

Sowing date

The date of sowing largely determines the state of maturity of the crop when it is utilised and should be adjusted accordingly. Although early sowing usually gives the highest yields, sowing too early in relation to the date of utilisation results in an excessively mature crop; the stems become very woody, with consequently heavy

wastage. The earliest sowings made in late March, April and May
are suitable for grazing in August and September but most farmers
wisely prefer to delay the main sowings for autumn use until June
or perhaps even the first week in July. Sowing much later than mid-
July in the south of England and after early June elsewhere is
usually inadvisable, as the period of warm weather left for growth is
normally too short to produce a worthwhile crop before winter sets
in.

Plant density and seed rate

Marrow stem kale requires a fairly high plant density. Sowing too
thinly can result in relatively sparse crops, which then make utilisation exceedingly difficult. At low plant densities the stems become
excessively thick, coarse and less acceptable to cattle; such crops are
also particularly prone to lodging and to stem cracking (Plate 1),
when they split longitudinally and expose the pith, which may subsequently rot. The erstwhile practice of singling kale is therefore
highly undesirable. Most commercial varieties of marrow stem kale
are rather susceptible to lodging, especially after heavy applications
of nitrogen.

With early sowing before mid-June in the South, row width and
spacing within the row do not greatly affect the yield of dry matter,
provided that very low plant densities are avoided. Row widths of
up to 20 inches are usually satisfactory. When late sowing occurs
after mid-June, row width should not exceed 14 inches; a row width
of 7 or 8 inches is often preferable.

The seed rate usually varies from 2–4 lb per acre according to the
seeding mechanism of the drill, the time and the method of sowing.
Except with a precision drill, when a 2-inch spacing using $1–1\frac{1}{2}$ lb
seed per acre is satisfactory, much less than 2 lb seed per acre may
well give a thin stand and prove to be false economy. The higher
rate, 3–4 lb per acre, is preferable for direct drilling or where seed
distribution in the row is likely to be uneven.

Hybrid Varieties

The hybrid varieties Maris Kestrel and Proteor (Plate 2), although
rather similar in appearance to marrow stem kale, are markedly
superior in several respects and illustrate the extent of the improvement the scientific plant breeder can achieve.

Both varieties are much more frost hardy than marrow stem,
although not as hardy as the best dwarf thousand head kales and are
more suitable than marrow stem for folding after Christmas. Maris
Kestrel gives a rather better yield of digestible dry matter than Proteor, and the best stocks of marrow stem kale; the stems of both

hybrid varieties are shorter and much less susceptible to lodging than in most stocks of marrow stem.

Owing to the much thinner ring of conducting tissue around the pith and their higher soluble dry matter content (sugars, etc.), the stems of the hybrid varieties are more digestible than those of marrow stem kale; Maris Kestrel is outstanding in this respect. Maris Kestrel is also of obvious value for extending the grazing season of the marrow stem type of kale right up to March. The greatly improved performance of this variety fully justifies the small additional cost of the seed. The hybrids may be sown as late as marrow stem kale but the seedrate may be reduced to 2 lb per acre.

Thousand Head Kale

The yield of stem and total green yield of thousand head kale is considerably lower than that of marrow stem and the hybrid varieties of kale; well grown crops usually give a total yield of 15–20 tons per acre. The stem is relatively slender and unlike marrow stem kale, soon becomes very woody so that it is unpalatable, less digestible and largely neglected by stock; in assessing the yield of thousand head kale, the weight of stem should therefore largely be ignored.

The valuable part of this type of kale consists of the leaf and succulent young side shoots, which are mainly produced in late winter and early spring; a well-grown plant then presents a bushy appearance. Thousand head produces a much higher proportion of leaf to stem than marrow stem kale, although the precise ratio differs greatly from one type to another. Thousand head kale is also much more frost hardy than marrow stem. There are two distinct types:

(i) Normal.

(ii) Dwarf thousand head kale. (Plate 3)

The normal type has a relatively long stem (38–40 inches) which makes up 45–50 per cent of the total yield. The yield of stem is considerably higher than with the dwarf type, although there is little difference between the two in their yield of leaf. Hardiness is not as good as with the dwarf types. Flowering occurs later than in marrow stem kale.

The dwarf type, in which the stems vary from 20–28 inches in length, according to variety, is the hardiest of all the *B. oleracea* kales and flowers last. The proportion of leaf, which under suitable conditions, can reach about 70 per cent of the total yield, is much higher than with the normal type (50–55 per cent). Dwarf thousand head is also much less susceptible to lodging than the normal type; the variety Canson, which has the shortest stem of all, is outstanding in this respect.

As the yields of leaf and side shoot of the dwarf and normal types of thousand head are about the same, there seems to be little point in growing the normal type, particularly as the longer stumps which remain after feeding are more difficult to dispose of before ploughing.

Thousand head kale is useful for feeding sheep or cattle from January to early April, when a high degree of hardiness is required. It is often grown with swedes, either in mixture or in alternate rows; for instance, a five-row drill may sow one row of kale and four rows of swedes giving alternate strips of two rows of kale and eight rows of swedes. The proportion may be varied as required. Young kale side shoots are particularly palatable to young lambs and also easier on the teeth of the older ewes.

Thousand head kale appears to be resistant to the common races of clubroot.

Sowing date

Thousand head kale is usually sown during June and in early July, although in the milder areas of the south and west of England, sowing may occur up to the third week in July. Earlier sowing is unnecessary as it does not appear to increase the yield of side shoot in the spring and tends to result in extensive growth of stem, which may be difficult to dispose of before ploughing. If sown too late, there is insufficient time to allow adequate growth to occur before winter sets in. (Plate 3) A succession of sowings seems to have little practical value with this crop.

Plant density and seed rate

In contrast to marrow stem kale, thousand head needs adequate room in which to develop its side shoots; sowing thickly causes these to be almost totally suppressed (Plate 3). Unless wider rows are required to facilitate weed control, rows 14 inches apart are satisfactory and give a more even spacing of the plants. Where the seed can be evenly distributed with a precision drill or semi precision drill, $\frac{3}{4}$–1 lb of seed per acre should be adequate. With other drills and with direct drilling $1\frac{1}{2}$ lb seed appears to be ample. Where necessary, the seed should be bulked with 'seed size' tapioca; thick sowing must be avoided.

THE FORAGE RAPES

This group includes forage rape and the so-called swede-like or rape 'kales'.

Forage Rape

Forage rape belongs to the same species, *Brassica napus L.* as the swede. Rape is particularly valuable where speed of growth is the prime consideration, as it is capable of producing a large bulk of

PLATE I
Marrow Stem Kale

The effect of plant density on lodging and stem cracking. Experiment sown on 24th May, 1961 in rows 19in apart. Received 126 units N per acre. Photographs November, 1961.

► *...gh Plant Density:* four row ...ot of unsingled kale giving ...1 plants per yard of row; ...te thin stems standing well.

◄ *Low Plant Density:* four row plot of kale singled to mean spacing of 2 plants per yard of row, giving massive individual stems with severe lodging and stem cracking. Note collapsed plant in foreground, the result of the rotting of a cracked stem.

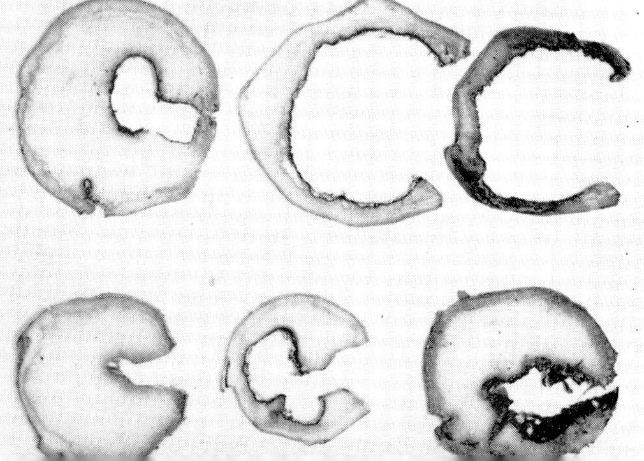

► *Stem Cracking:* transverse sections through cracked stems of marrow stem kale grown at wide spacings, showing various stages of decomposition of the pith. Bottom left: the incipient crack; top right: the pith completely gone.

PLATE 2
Marrow Stem and Hybrid Kales
The effects of variety and seed-rate. Photographs 19th November, 1971.

Left: Cannells marrow stem kale. Right: Maris Kestrel hybrid kale. Both varieties direct drilled at 4lb seed per acre on 2nd June, 1971.

Maris Kestrel hybrid kale. Left: seedrate 1lb per acre. Right: seedrate 4lb per acre. Both treatments direct drilled on 2nd June, 1971.

Hybrid kales. Left: Maris Kestrel (2 plants). Right: Proteor (2 plants). Supplied by NIAB from their trials at Seale-Hayne Agricultural College precision drilled on 12th July, 1971 at 2in spacing in rows 20in apart.

PLATE 3
Thousand Head Kale

The effects of variety, seed-rate and sowing date. Photographs 19th November, 1971.

Left: a normal type variety. Right: a dwarf type variety (Canson). Supplied by NIAB from their trials at Seale-Hayne Agricultural College precision drilled on 12th July, 1971 at 2in spacing in rows 20in apart.

Canson thousand head kale. Left: seedrate 1lb per acre. Right: seedrate 4lb per acre (2 plants). Both treatments direct drilled on 2nd June, 1971.

Canson thousand head kale. Left: direct drilled 2nd June, 1971. Right: direct drilled 27th August, 1971, seedrate 1lb per acre.

fodder in a very short growing season, when it will outyield the *B. oleracea* kales; with early sowing, however, the reverse is the case (Plate 4).

With good growing conditions forage rape can produce a fresh yield of around 12 tons per acre two to three months after sowing; in NIAB trials, yield has varied from 6–24 tons per acre. Forage rape is therefore very suitable for growing either as a catch crop, as a pioneer crop in the reclamation of marginal and hill land or for the provision of fodder in cases of emergency.

Providing that it is fed in a succulent stage of growth, forage rape is highly digestible, rich in protein, and exceedingly palatable to sheep. It is one of the best foods for fattening lambs in late summer and autumn, and is frequently grown specifically for this purpose.

Varieties introduced so far are not really frost hardy and are therefore unsuited for general use after Christmas. In the milder parts of the country, such as Devon, forage rape has often been used successfully for feeding breeding flocks of ewes lambing in the early months of the year. Unfortunately leaf losses can be heavy in frosty weather and the practice is very risky where severe winters occur.

Two types

Forage rape can be broadly divided into two types:

 (i) Giant rape.

(ii) Dwarf rape (Plate 5).

In practice, however, the two types form a continuous series, with no clear line of demarcation between them.

Giant rape is taller, with an erect habit of growth and a well-developed stem. It gives a much better yield than the dwarf type and is somewhat more palatable to stock, although the proportion of leaf is slightly lower. Giant rape should always be grown in preference to the dwarf type unless it is required to graze the regrowth.

Until fairly recently no clear varieties have been available and a number of different stocks of giant rape have been offered by merchants under such names as Early Giant, Giant English, or Essex Giant branching rape. The plant breeder has produced higher yielding named varieties such as Arvor, Bishop, Emerald, Fora and Lair and these are now available commercially.

Dwarf varieties of rape are relatively shorter and show little development of the stem. Dwarf rape is said to recover better from grazing, although this attribute is generally of somewhat doubtful value. The dwarf rapes have a slightly higher digestibility but are the least palatable. The standard variety has been Broadleaved Essex, offered as a number of different stocks of seed. Both giant and

C

dwarf rape are capable of producing side branches in the winter and early spring if the plants are not sown too thickly.

Fertiliser requirements

Forage rape is a particularly useful and versatile crop but one which has sometimes given very disappointing results. To succeed it must have ample soil moisture; it is unsuited to very dry situations. Satisfactory growth requires a liberal supply of fertiliser, especially nitrogen. To attempt to grow it on stubble without nitrogen is the height of futility and is virtually a guarantee of crop failure (Plate 5).

The sheer usefulness of rape often leads to the temptation to over-crop—resulting in an infestation of clubroot. Unlike the *B. oleracea* kales, most varieties of rape currently available are highly susceptible; infection results in collapse of the plants and total loss of crop. Varieties which are either tolerant or resistant have now been bred and are currently in trial.

The variety Nevin, bred at the Welsh Plant Breeding Station, is available commercially but the yield of this variety is lower than that obtainable from the taller varieties and it should be grown only where clubroot is likely to be a problem. Nevin can become heavily infected with clubroot but the crop does not appear to collapse and remains edible. The growing of such varieties can, of course, perpetuate clubroot.

Sowing date

Sowing can occur at any time between May and early September, but the date of sowing should be adjusted according to the date the crop is required for use and the area in which it is grown. Unless specifically required for early use, rape should not be sown very early as it is likely to be ready too soon and will have become unpalatable before it is required. Early sowings are also more susceptible to attacks by aphid, swede midge and mildew, which causes premature loss of leaf and palatability. Sowing too late gives only a light crop.

In the North and on the hill lands of Wales, sowing in June is generally best and sowing should not generally be contemplated much after the end of July. In the south of England rape is almost invariably grown as a catch crop, when a clean weed-free winter barley stubble provides an ideal situation. Sowings required for August use may be made in early June, but July and the first three weeks of August is the most usual period for sowing.

In the mildest parts of the South-West sowing can occur as late as the first week in September, a common practice on some coastal farms. However, these very late sowings are too risky for general use and the longer the delay in sowing, the greater the decline in the yield of dry matter.

Plant density and seed rate

Fairly thick sowing generally gives the most rapid ground cover and the highest yield of dry matter. A high plant density is particularly important with the later sowings, when time for full development of the individual plant is strictly limited; more plants are then needed to give a full yield of dry matter. Thick sowing also results in thinner stems which, when young, are more readily consumed by sheep. The seed may either be broadcast at 6–7 lb per acre or sown in rows seven inches apart with all coulters of a corn drill, using a seed-rate of 4–5 lb per acre. The cash saving achieved by sowing thinly is not worthwhile as the seed of all varieties is relatively cheap.

The Swede-Like or Rape 'Kales'

The group of swede-like or rape kales consists of two fairly well known varieties, Hungry Gap kale and Rape kale. These varieties, which may be distinguished by the curled margins of their leaves, are more frost hardy than ordinary forage rape; they also flower later. Their purpose is to combine rapid growth from a relatively late summer sowing in the latter part of July or early August with a fair degree of frost hardiness and to fill the 'hungry gap' which occurs before the grass starts growth in the spring.

In the south of England these varieties are best sown in July and August, the optimal date depending on the district. Early sowing should generally be avoided, as these varieties, like ordinary forage rape, are susceptible to powdery mildew which can cause severe loss of leaf. As dwarf thousand head kale is hardier and gives a higher yield from earlier sowings, these varieties should only replace thousand head kale when sowing occurs too late for the latter to give a reasonable crop (Plate 6). A seed-rate of 3–4 lb per acre drilled in 7-inch rows is adequate.

Hungry gap kale is rather less hardy than rape kale but will make very rapid growth from sowings made in July and August, giving a yield in autumn similar to that of giant forage rape. Unfortunately heavy loss of leaf occurs with the onset of sharp frost in early winter.

Except under very mild conditions, side branches are produced rather late, so that little new growth is available before the latter part of March. Useful keep can be obtained from this variety either in the autumn, when it may be treated like forage rape or alternatively it may be left until late March or April. This variety is also grown for human consumption.

Rape 'kale' is rather hardier than Hungry gap kale but gives a slightly lower yield; it is also said to show better regeneration after light grazing. This variety is only really suitable for growing to feed in late March or April. Giant rape will give a much better yield in the autumn.

FODDER RADISH

The outstanding feature of fodder radish is its capacity for making exceedingly rapid growth, which has in some cases far exceeded that of giant rape. The seed is three times the size of rape and the seedlings move quickly through the phases of germination and establishment, achieving a ground cover so quickly that weeds have difficulty in competing effectively. Fresh yields of up to 30 tons per acre have been obtained but 15–20 tons is more usual; the yield is generally heavier than that of rape over an 80–100 day growing period and fodder radish is often the more palatable of the two.

Varieties

Varieties have been classified by the NIAB as follows:

Early flowering type: Rapide; Siletta.

Late flowering type: Ardèche or Raifort Champêtre; Slobolt.

Extra late flowering type: Hailstone (White turnip rooted).

In the first two types the stems, which together with the light green hairy leaves, produce the bulk of the utilisable crop, attain a height of about $2\frac{1}{2}$ feet before the onset of flowering; the early varieties flower some 6–8 weeks after sowing, while the late varieties flower approximately three weeks later, about three months after sowing.

Varieties of the early type show little root development; those of the late type show rather more, especially at low plant densities when the individual roots may reach weights of up to 2 lb each and are readily eaten by sheep. However, the main function of these types is to produce tops, and at normal seed-rates, 8 lb per acre for 12–15 in rows and 10–12 lb per acre for 7 in rows and broadcasting, the roots make relatively little contribution to the total yield.

The major disadvantage with these two types of fodder radish, especially the early flowering type, is the shortness of the period when the crop is suitable for utilisation and at the same time combines a good yield with satisfactory digestibility.

The feeding value and palatability of the crop deteriorates rapidly with the onset of flowering (Plate 7) and the stems are soon neglected by stock. It is therefore necessary to organise a succession of sowings to provide a continuous supply of fodder which is in a suitable stage of development for consumption.

The extra late variety, 'Hailstone', does not usually flower after midsummer sowings and exhibits a biennial habit with a root development comparable to that of a white turnip but with a larger proportion below ground level (Plate 7). The absence of flowering and lack of deterioration in feeding value of this variety shows to advantage over the earlier flowering types, but its yield is considerably lower than that of other varieties of fodder radish and also of forage rape.

Sowing instructions

The dual pitfalls of sowing too early and of sowing too large an area at one time should be avoided. Sowing generally occurs from mid-July to mid-August, according to the expected date of utilisation. Except in the milder areas of the south-west of England, sowings made after mid-August are risky and failures have occurred.

The crop should be fed in early autumn rather than in early winter, as fodder radish is very susceptible to frost damage. It has been observed that the roots of the 'Hailstone' variety possess a useful degree of frost resistance and keep better than most white turnips; the tops, however, are susceptible to frost damage.

Fodder radish appears to be rather more sensitive to adverse conditions in the seedbed, such as cold, drought and excess moisture than is rape. The crop is also less suited to the colder areas, as in the north of England, although it has proved to be quite useful on some exposed hill farms in Wales. The great advantage of rape over fodder radish is that its biennial habit, combined with absence of flowering in the year of sowing and somewhat better frost tolerance, allow it to be utilised satisfactorily over a much longer period.

Suitability of crop

The capacity for very quick growth makes fodder radish very suitable for use as a pioneer crop in land reclamation and growing as a catch crop. It is particularly useful for fattening lambs and has also been used for dairy cows. Fodder radish does not appear to be affected by clubroot, even under conditions where rape has been completely wiped out, which is of obvious advantage on clubroot-infested fields. Fodder radish is also unaffected by powdery mildew; neither, apparently, does it increase beet eelworm populations.

In spite of the considerable publicity and interest when first introduced into the United Kingdom, fodder radish has never achieved a position of any importance. The very short period during which the crop remains usable once ready and its lack of frost hardiness have weighed heavily against it. This is a pity, because its rapid growth, heavy yield, palatability and disease resistance are valuable attributes.

Although Slobolt is a marked improvement on the earlier flowering sorts, varieties which are very slow to flower and yet retain the capacity of the crop to make rapid growth are still needed; new and successful varieties could increase the practical value of this crop appreciably.

CATTLE CABBAGE

The main type of cabbage cultivated specially for cattle feed is the Devon Flat Poll (Plate 6). Under conditions of very high fertility yields of up to 60 tons per acre may be obtained but 25–35 tons per acre is more common. The crop is normally used for feeding to cattle from September to mid-December. Probably the only real advantage of this crop is the relative ease with which it can be carted to livestock. Although the fresh yield is very good, its dry matter content, approximately ten per cent, is low and comparable crops of kale will give a higher yield of dry matter per acre. The crude protein content and the digestibility (D value about 71) are both very satisfactory. Once fully mature, this crop does not stand frost well and is inclined to rot. Crops with a high proportion of loose 'rosette' heads are difficult to handle.

When to sow
Flat Poll cabbage plants are normally raised in a plant bed and transplanted in April or early May into their final cropping position. Very often farmers buy the plants from specialists; others raise the plants from seed themselves. The seed is usually sown in a sheltered seedbed in August or early September in mild areas. One pound of seed, sown on 1/10 acre in rows 9–12 inches apart, is usually adequate to provide plants for 1 acre of crop. In the field the rows vary from 30–36 inches apart and a spacing of 36 inches is allowed within the row, requiring approximately 5,000–7,000 plants per acre.

The residues of cabbages grown for human consumption may also be used very profitably for feeding to livestock, particularly such hardy types as January King which provide food in the difficult months in the new year. The advantage of these and such crops as Brussels sprouts is that they provide a substantial cash income as well as useful fodder for stock, especially for sheep grazed in situ. However, before embarking on this type of cropping, it is essential to have obtained an assured market.

Although cabbages grown specifically for fodder are capable of providing a large amount of food from a small acreage, which is obviously of considerable importance where shortage of land is a prime consideration, they tend to have a relatively high labour requirement and occupy the land for a large part of the year.

A number of the other fodder crops such as kale and rape give a relatively better return in a much shorter period of time and yet allow a cash return or a crop of hay or silage to be obtained from the land on which they are grown. In these circumstances the small area of cabbage now grown specifically for fodder in England and Wales would appear to be a fair comment on the prospects of this crop!

MANURING OF THE BRASSICA GREEN FORAGE CROPS

As acid soil encourages the development of clubroot, it is the height of folly to attempt to grow brassica crops under acid conditions; lime should always be applied where necessary. The precise pH that is desirable depends on soil type: with very light soils a pH 6·0–6·5 is optimal, whereas on the somewhat heavier soils a pH of 6·5–7·0 is preferable.

The forage brassicas are gross feeders and liberal manuring, particularly with nitrogen, is essential. When brassicas are grown as maincrops they are generally adequately manured but all too often farmers growing these crops as catch crops apply little or no fertiliser in an effort to cut costs to the bare minimum; the result is almost invariably half a crop or even total crop failure.

Crops are often under-manured in the belief that there are ample manurial residues from the previous crop; frequently this is not the case. Heavily manured crops such as early potatoes do leave substantial residues behind but there is precious little left after worn out permanent pasture or a cereal crop. In the latter case phosphate and potash residues may be adequate but liberal nitrogen application is still essential.

As well as being influenced by the previous crop, its management and the farming system, the nitrogen requirements of the brassica green forage crops depend in the length of the available growing season, as influenced by the dates of sowing and consumption of the crop. The longer the growing season, the better the response is likely to be to heavier application of nitrogen. The nitrogen application should also be restricted when crops are to be left to stand well into the new year.

Over-generous application of nitrogen tends to produce soft sappy growth which is much more susceptible to frost damage. At the same time it is essential to strike a reasonable balance and to give enough nitrogen to secure adequate growth of the crop. Recommendations for the nitrogenous manuring of the brassica green forage crops are given in Table 3.

Flat poll cabbages merit separate consideration; like mangolds they are expensive to grow and generous treatment is essential to obtain the massive yield necessary to make the crop worthwhile. Cabbages should always be grown on fertile ground and, if at all possible, a really heavy dressing of well rotted farmyard manure should be given, especially after a cereal crop. Nitrogen should also be generously applied (Table 3).

TABLE 3. Recommendations for the Nitrogenous Manuring of the Brassica Green Forage Crops Grown with Conventional Cultivations*

CABBAGES AND KALES

	Units N per acre[1]		
Previous crop	Flat poll cabbage	Kale sown before mid June	Kale sown after mid June
Permanent grass or ley grazed and in good heart. Sward usually with high clover content prior to ploughing.	80–100[2]	60	40
Poor old permanent grass or ley cut for hay or silage	110–140[2]	100	80
Cereal grown in previous year (followed by maincrop cabbage/kale)	130–160[2]	100	—
Cereal or cereal silage (followed by catchcrop kale)	—	—	80–100
Heavily manured early potatoes	—	0–40	0–40

[1] For thousand head kale and other kales to be eaten after Christmas reduce dressing by up to 20 units.

[2] Use lower amount of nitrogen for crops receiving *heavy* application of FYM.

RAPE AND FODDER RADISH
(use higher rates for fodder radish, either for rape)

Previous crop	Units N per acre
Permanent grass or ley grazed and in good heart. Usually with high clover content prior to ploughing	40
Poor old permanent grass or ley cut for hay or silage or grown as pioneer crop	60–80
Cereal (catch crop sown July/early August)	60–80
Cereal (catch crop sown late August—S. England)	60
Heavily manured early potatoes	0–40

*Recommendations for direct drilled crops are given in the appropriate chapter; these are generally about 20 units per acre *above* those for conventionally drilled crops. Sandy soils generally need rather more generous treatment and up to 20 units N per acre *extra* may be given for the earlier sowings.

Plant bed (cabbages)

When the plants are to be produced at home the plant bed should be on clean land on which brassicae have not been grown for at least four years. The land must not be acid and should usually be limed liberally during the preparatory cultivations as a precaution against clubroot.

The plants must not make soft sappy growth: they must be well grown and yet at the same time in 'hard' condition when required for planting out the following spring. The level of nitrogen application should therefore be enough for adequate growth and no more. The following seedbed dressings are suitable:

Moderately Fertile Soils	Highly Fertile Soils
40 units N	20 units N
80 units P_2O_5	60 units P_2O_5
80 units K_2O	60 units K_2O

Recommendations for the application of phosphate and potash, which depend largely on the previous manuring and soil type of the field, are given in Table 4. These tables can only constitute a general guide and in some circumstances it may be necessary to apply quantities outside those recommended. All fertiliser should be broadcast and worked into the seedbed during seedbed preparation.

Placing concentrated fertilisers in contact with the seed is dangerous and should be avoided, although up to 1 cwt of superphosphate may be used if it is necessary to use a fertiliser to bulk the seed.

Only nitrogen should be topdressed and then only when absolutely essential. Late top dressings should be avoided as these have been known to produce toxic effects in the stock consuming the crop.

TABLE 4. Recommendations for the Phosphate and Potash Manuring of the Brassica Green Forage Crops

CABBAGES AND KALES

	Units per acre	
	P_2O_5	K_2O
Standard dressing	100	100
Low soil P_2O_5* or K_2O—increase either up to	125	125
Good soil reserves of P_2O_5 or K_2O	50	50

RAPE AND FODDER RADISH

	Units per acre	
	P_2O_5	K_2O
Standard dressing	30–50	30–50
(give the heavier dressing where crop yield comparable to kale is expected)		
Low soil P_2O_5* or K_2O—increase either up to	75–100	75–100
Good soil reserves of P_2O_5 and K_2O or when grown as stubble catch crop on soil well manured with P_2O_5 and K_2O	0–30	0–30

*On newly reclaimed land or in high rainfall areas 10 cwt per acre high grade basic slag should be applied which may replace approximately 70 units of P_2O_5 in the recommendation. Additional potash may be applied as 60 per cent muriate of potash if required.

Chapter 3

THE FODDER ROOT CROPS

SWEDES, TURNIPS and mangolds are the main root crops grown for fodder in Great Britain. Fodder beet is quite unimportant. The root crops are capable of giving very high yields of dry matter per acre, are palatable to stock and retain their high digestibility over a long period. Average yields obtained in NIAB trials in England and Wales are shown in Table 5.

TABLE 5. Average Yields of Root Crops Obtained in NIAB Trials in Recent Years

| Crop | Situation of Trials* | Yield in Tons/Acre | | Dry matter content % |
		Dry matter	Roots	
Swedes	North and West	2·5	27	9·3
Turnips White-fleshed	North and West	1·9	27	7·2
Yellow-fleshed	North and West	2·0	25	8·0
Mangolds	South	4·9	47	10·4
	North	3·3	29	11·2
Fodder Beet	South	5·5	33	16·7
	North	3·9	22	18·0

*North=north of 53°N.   South=south of 53°N.   West=west of 2°W.
Reproduced from NIAB Farmers Leaflet No. 6 1972 "Fodder Root Crops".

Swedes and turnips are particularly suited to the north and west of England and Scotland. Yields in Scotland and the north are generally much heavier than in the south, where 15-20 tons per acre is a normal crop. In the northern areas, crops of 35 tons per acre and over are commonplace. A 40-ton per acre crop of swedes at 10 per cent dry matter will give 4 tons of dry matter per acre.

While yields of different types of mangold and fodder beet vary considerably, crops giving widely divergent tonnages of fresh roots may give an identical yield of dry matter. Thus a 30-ton crop with a 10 per cent dry matter content gives an identical yield of dry matter to a 20-ton crop with a 15 per cent dry matter content (i.e. 3 tons dry matter per acre). In the latter case the food is merely more concentrated. The choice depends on the type of stock for which the crop is required. The high dry matter roots can be very hard on the teeth, particularly on those of sheep and young cattle.

SWEDES AND TURNIPS

Swedes may be readily distinguished from turnips by the presence of a 'neck' from which the leaves arise; the latter are smooth and usually bluish or dark green. In the turnip the 'neck' is so small that it is almost or completely absent and the leaves are coarsely hairy (Plate 8).

Swedes have a higher dry matter content and consequently a better feeding value than turnips. They are also capable of giving a markedly higher yield of dry matter per acre and are, with the exception of a few varieties, much hardier than the white and soft yellow turnips. Swedes are therefore better adapted to feeding later in the winter.

In Scotland and the northern areas of England, feeding may start in November. Elsewhere they are usually fed from January to early April, according to variety. In areas where very cold winters are experienced all but the very hardiest varieties are clamped or stored under cover, particularly if fed in February or later. In other areas, they are normally hardy enough to be fed in situ.

The proportion of a crop that is fed as required and the proportion that is clamped varies greatly from one district to another. There is no set pattern. There appears to be little place for varieties of swedes which are not frost hardy. Swedes must be fed before the flowering stem starts to grow in the spring or the food material stored therein is withdrawn, leaving the roots spongy and unpalatable. There are very large varietal differences in dry matter content, time of maturity and frost hardiness. Varieties may be divided into three major groups on the basis of skin colour:

Purple skinned varieties
These are the most widely grown. This group may be subdivided into two major groups, which have light and dark purple skins, although there is no clear cut line of demarcation; the two groups form a continuous series. Skin colour tends to be darker in small topped varieties and is also affected by conditions of growth and sunlight. The shading effect of large tops sometimes produces a less intense skin colour. Light skinned varieties give good yields of roots but with a lower dry matter content and are less hardy; they are often unsuited to feeding in situ, especially where there is a risk of very

cold winter conditions. The dark skinned varieties are considerably hardier.

A further group, the 'Hardy Purple skin' or 'North Type' of swede is grown in the north of Scotland, to where it is largely confined. These varieties are very late maturing, extremely hardy and if left in drills and furrowed up will keep well until late May. The type includes varieties such as Aberdeen Prize, Balmoral, Buchan and Coxton Crofter. This type of swede is quite unsuited to English or Welsh conditions.

Bronze skin varieties have traditionally been considered to possess characteristics which are intermediate between the dark purple and green topped types. In point of fact they show a wide range in their dry matter contents, degree of winter hardiness and times of maturity.

Green skin varieties tend to grow somewhat smaller roots than the other two types but provided spacing is close enough, produce a very satisfactory yield of dry matter. In the United Kingdom green skinned swedes are principally grown in north east Scotland. This group includes some of the latest maturing and hardiest varieties in cultivation. They keep well in ground and store and are ideally suited to late feeding. In cold areas the latest varieties can be stored and fed well into April.

Colour of flesh

The colour of the flesh of the mature root may vary from yellow to white. Practically all varieties in current cultivation have yellow flesh; the only white fleshed variety of current importance is Wye.

Field characteristics

Greater winter hardiness and later maturity are associated with increasing dry matter content of the roots. In some of the hardiest varieties the skin is tough and the flesh is hard, which can ruin ewes' teeth. Varieties with improved yield and dry matter content but without the associated hardness of flesh are desirable. NIAB trials indicate that there is considerable overlap in the dry matter content of the various types of swede; skin colour and dry matter content cannot therefore be linked together automatically. In England and Wales dry matter content usually varies from 8–11 per cent. The analysis of swedes has been shown to be affected by the latitude in which they are grown. The slower growing season of the northern areas favours the production of a higher dry matter content, higher soluble carbohydrate and delay in lignification. In northern Scotland some of the hardiest types of swede may attain a dry matter content of 12 or even 13 per cent.

Root shape can be of considerable importance. When swedes are grown for human consumption, as in Devon, fairly small uniform globe shaped roots of about 2 lb each with a bright skin and an extremely small short neck are required. A very small neck is of considerable importance because a heavy neck leaves a large scar when the swede is topped, detracting from the appearance of the roots and tending to increase the loss of weight due to evaporation in transit to the shops.

Excessive nitrogen is known to increase size of neck. Coarse or fangy roots are highly undesirable. The variety Acme, with its small neck and extremely uniform root, is particularly suitable and widely used for the market trade. Seed is best obtained straight from the breeder. Devon Champion is also popular.

In the north of England swedes for the market are usually taken from commercial crops; very often half swedes are sold. Uniformity of root size and shape is also important for machine harvesting, where a globe or flat globe is preferable. The neck should be of moderate length; excessively large necks are inclined to give difficulties with harvesting. Pentland Harvester, which was specially bred for the purpose, is particularly suitable.

When grown for folding the roots should sit well on top of the ground. Tankards or intermediates may provide the best shape for this purpose but Globe varieties seem to be somewhat hardier.

Disease resistance

Disease resistance is extremely important. No variety yet available is completely resistant to powdery mildew (Erysiphe polygoni) but Chignecto, Bangholm Sahna, Bangholm, Bangholm Wilby, Ruta Øtofte, Vogesa and most of the green topped varieties are the least susceptible to infection. Varieties of swedes resistant to clubroot, Plasmodiophora brassicae Wor, are listed in Table 11.

Stocks from the following varieties are recommended by the NIAB for use in England and Wales:

Purple Skin—Globe shape: Acme, Bangholm Sahna, Chignecto, Pentland Harvester. **Intermediate shape**: Bangholm, Bangholm Wilby, Danestone 1, Doon Major, Magnificent, Ruta Øtofte, Victory.

Bronze Skin—Globe shape: Ne Plus Ultra. **Intermediate shape**: Vogesa.

Green Skin—Globe shape: Wilhelmsburger, Wilhelmsburger (Sator) Øtofte. **Intermediate shape**: Dalo, Danila.

Full details of the above varieties and the names of the seedsmen offering the recommended stocks of seed may be found in Farmers' Leaflet No. 6, National Institute of Agricultural Botany, Huntingdon Road, Cambridge.

Types and varieties of turnip

Turnips mature more quickly than swedes, give a higher yield in a shorter growing season and can be sown much later, although they exhibit a lower dry matter content and feeding value. Types of turnip differ widely in their time of maturity, keeping quality and degree of frost hardiness. The crop may be grown either for the production of roots or tops.

Turnips may be divided by the flesh colour of their roots into two distinct groups of White and Yellow turnips.

White turnips

This type makes the most rapid growth but has the lowest dry matter content and feeding value. White turnips cannot be regarded as hardy and should be fed before severe winter weather sets in. They are particularly suited to catch cropping and to later sowings, where a good yield is wanted quickly. White turnips are also suitable for the production of tops when they can be sown much later than when required for roots. Varieties of white turnip may be divided into three groups:

Autumn stubble turnips

A considerable number of new varieties of autumn stubble turnips, mostly of Dutch origin, have been introduced in recent years. Many more are currently in trial. The tops are larger than the traditional English types and the roots are usually cylindrical in shape (Plate 8). These turnips make exceedingly rapid growth and will produce surprisingly good yields in a very short space of time. Sown at Seale-Hayne on July 25th, 1970, after winter barley, Debra turnips gave a yield of $17\frac{1}{2}$ tons of tops and roots per acre by late November. Even when sown as late as September 3rd, 1970, Debra, Labra and Vobra gave yields of between 9 and 10 tons of tops per acre. However, this type of turnip must be fed while still growing actively and while the bulbs are still palatable. If they are left too long the roots become woolly or even hollow and are totally neglected by stock. A number of these varieties show excellent resistance to clubroot (Table 11). Because of their more rapid growth and resistance to clubroot these varieties should usually replace the traditional varieties for growing as stubble catch crops—they appear to be markedly superior for the purpose.

'Traditional' white fleshed varieties

Traditional white fleshed varieties are mainly used to produce a heavy crop of bulbs for feeding and are either globe or tankard in shape. The following varieties are well known and offered by various seedsmen—

Green or Imperial Green Globe: Best general purpose variety in this group. Keeps well for a white turnip.

Lincolnshire Red Globe: An early heavy cropping variety.

Pomeranian White Globe: An early variety which makes rapid growth and large roots. Will not keep.

Purple Top Mammoth: One of the best early varieties, producing the largest roots.

While these varieties are suited to late sowing they are not hardy and must be used before winter.

Flat hardy green round turnips

Flat hardy green round turnips are grown mainly for tops in either pioneer crops, catch crops or mixed with Italian ryegrass where a hardy type of turnip is required. They have a higher dry matter content but give a much lower yield of roots than other varieties and are suitable only when a later sown turnip is required to stand the winter.

Yellow turnips

Yellow turnips fall into two groups—

Early yellow fleshed or soft yellow turnips: These turnips are quick growing and relatively high yielding but are slightly hardier than the white turnip. They do not generally keep unless well protected from frost and have a low dry matter content, usually between 7–8 per cent.

Varieties—Globes: Centenary, Fosterton or Dales Hybrid, Grampian, Sheepfold. **Tankards:** Bortfelder, Yellow Tankard.

Hardy Yellow Turnips or Yellow Turnips: Agriculturally (but not in appearance) this type is closer in its behaviour to swedes than to white turnips. Hardy yellow turnips generally have the highest dry matter content among the turnips, 8–9 per cent and are hardier and better keepers than the white and soft yellow types but are slower growing and later in reaching maturity. They are hardier than some varieties of swedes. While hardy yellow turnips have to be sown earlier than the two previous groups they can be left in the ground much longer. In the south of England, hardy yellow turnips can be left in the ground all winter, while even in the North, the Wallace and sometimes the Bruce can be left out in some areas.

Varieties—Green Top Globes: Aberdeen Green Top, Balmoral, Challenger Glenlogie, Perfection, The Wallace. **Purple Top Globes:** Aberdeen Purple Top, The Bruce.

The Bruce and the Wallace are worthy of special mention. The Bruce is probably the heaviest yielding variety of yellow turnip in cultivation. With a dry matter content of around 10 per cent this variety can give a higher yield of dry matter than some varieties of swedes, especially in upland areas of Scotland where the growing season is shorter and the soil is often thinner. The Wallace has a comparable dry matter content and the bulbs keep better but the yield is somewhat lower. Selected stocks of both of these varieties have shown a good resistance to at least some common races of clubroot, although the Bruce is said to be the more resistant of the two.

Yellow turnips are not widely grown nowadays and it appears that their limited use is largely confined to Scotland. Elsewhere, as on some of the higher ground in south west England, some farmers still grow yellow turnips to obtain hardy roots to feed sheep after Christmas. In the south, yellow turnips can be sown after a cut of silage or hay and the author has grown them after an early harvested crop of winter barley in Devon, although maximum yields can then hardly be expected.

A technique requiring the minimum amount of labour has been developed for growing and managing a mixture of Italian ryegrass and hardy yellow turnips for sheep keep. The technique is based on the fact that the hardy yellow turnip is not palatable to sheep until it matures late in the autumn.

The mixture ($\frac{1}{2}$ lb Aberdeen green top yellow turnip+20—25 lb Italian ryegrass of a hardy variety) is sown in June or early July, according to the district. With appropriate management the ryegrass may be grazed in the late summer and autumn, followed by the turnips during the winter. The ryegrass provides early bite the following spring.

The crop is grazed as soon as there is enough ryegrass to hold the sheep. At this stage it is desirable to graze the ryegrass fairly hard to prevent it from becoming unpalatable or from competing with the turnips. After prolonged hard grazing the sheep may eat some of the turnip leaves; a few days rest usually cures the trouble. The grazing of the ryegrass is continued until the turnip bulbs become palatable, which in Devon occurs about mid-November. The stock are then removed for about six weeks to allow the turnips to complete their growth. Grazing is then resumed and the turnips are folded during the winter; the area of the fold should be small, as turnips lose their frost hardiness once the bulb has been bitten open.

Once the turnips have been cleared the ryegrass may be top-dressed to produce spring grass. The land may then be reseeded or if the ryegrass is still highly productive the sward may be retained until it is ploughed for a cereal crop the following autumn. Once the turnips are cleared, the Italian ryegrass should receive generous applications of nitrogen throughout the rest of its life.

Sowing dates of swedes and turnips

In Scotland and the north of England swedes are grown as a 'main' crop and it is necessary to obtain the heaviest possible yield, 30 tons per acre or more. A long growing season is required and early sowing is therefore absolutely essential. Swedes are generally sown in these northern areas from the last week in April until about May 25th. If possible sowing should be completed by the second week in May, when the maximum yields of roots may be obtained (Table 6); delay beyond this date usually reduces the yield of roots considerably.

TABLE 6. The Effect of Sowing Date on the Yield of Swedes and Yellow Turnips (Extracted from Advisory Bulletin No. 6—Turnips and Swedes. Department of Agriculture and Fisheries for Scotland).

SWEDE TURNIPS: Mean yields in tons per acre over 10 years.

Sowing Date	Fresh Weight (tons)	Dry Matter (tons)
1st May	24·8	3·1
8th May	24·3	3·0
15th May	23·2	2·9
22nd May	21·4	2·6
29th May	19·9	2·5
5th June	18·1	2·3

In 7 out of 10 years the highest yields occurred in crops sown on 1st May or 8th May.

YELLOW TURNIPS: Mean yields in tons per acre over 10 years.

Sowing Date	Fresh Weight (tons)	Dry Matter (tons)
8th May	22·2	2·4
15th May	22·0	2·3
22nd May	20·9	2·3
29th May	19·9	2·2
5th June	18·0	1·9
12th June	17·0	1·8

In 6 out of 10 years the highest yields occurred in crops sown on 8th May.

Green topped varieties of swede should be sown before mid-May. However sowing should not occur before the cessation of heavy morning frosts, as these may cause a marked increase in the number of bolters (plants which run to seed prematurely in the year of sowing; turnips and swedes are biennials and do not normally flower and produce seed until the year following sowing). Late sown swedes tend to produce excessive foliage and little root. If swedes have not been sown in Scotland or the north of England by the end of May, they should be replaced by yellow turnips, which will then give a higher yield of dry matter per acre.

Hardy yellow turnips require early sowing, which should generally be completed by the end of May, particularly in the coldest northern

D

areas. As with swedes, the earliest sowings usually give the best yields (Table 6). Early yellow fleshed or soft yellow turnips may be sown a fortnight later, usually up to the middle of June. Thereafter, only white turnips or continental stubble turnips should be sown.

Swedes and turnips may be sown in the south of England about a month later than in Scotland and the north of England. In contrast to the cooler northern areas, early sowing in the South not only shows no advantage in terms of yield but renders both swedes and turnips much more liable to attacks of powdery mildew, especially where good early progress of the crop is subsequently arrested by hot dry weather. Swedes may generally be sown between the last week in May and early June, although the first three weeks of June is probably the optimum period.

In Devon only swedes required for early marketing as a vegetable are sown in May; the bulk of the crop is sown in June; in fact many growers prefer to wait until the traditional optimum period around June 21st to 24th. However, the latter dates should be regarded as the latest time of sowing and not as an optimum if a full crop is to be achieved. Sowings of swedes made up to mid-July have frequently been known to produce useful crops but are risky and the yield of roots can be seriously reduced if germination is delayed by dry conditions.

In hill areas it is preferable to complete sowing before mid-June. If turnips are required for the production of roots or bulbs in the south of England or Midlands they should be sown in June and July according to the district.

As a general guide, the sowing of hardy and soft yellow turnips should be completed by about the end of June and mid-July respectively. Thereafter white turnips should be sown.

White turnips are now generally grown as a catch crop after a winter sown cereal harvested in late July or early August and are hardly ever grown as a traditional root crop.

White turnips can generally be sown in the southern half of England up to the end of August and even up to the end of the first week of September in the mildest areas, although the total yield of roots declines sharply as August progresses. Sowings made after mid-August give a small yield of roots, but a good yield of tops can be extremely useful.

The later sowings show improved frost resistance and if required will stand a mild to moderate winter quite well; the green feed is particularly useful to ewes immediately after lambing. Continental stubble turnips are particularly satisfactory where the crop is to be consumed before the advent of hard weather but they usually lack hardiness and are best replaced by hardy green round turnips where there is a likelihood of the crop being left to stand the winter.

Plant population and seedrate

Plant population within the range of approximately 14,000 to 53,000 plants per acre appear to have little practical effect on the total yields of roots and dry matter produced by swedes grown on fertile soils; only when the number of plants falls below this level— spaced 20 inches apart in 22-inch rows—is the total yield of roots at all adversely affected.

Even with very irregular spacing at populations as low as 15,000 plants per acre, the loss of yield is likely to be small. At normal farm spacings therefore, plant population is likely to be relatively un-important as far as the total yield of swedes is concerned. However, closer spacing does give a modest increase in the yield of tops which is of obvious practical importance if the crop is to be fed in situ before the bulk of the leaves are destroyed by frost. Soils of lower fertility require higher plant population.

The presence of plant disease alters the situation completely. Low plant populations have been shown to increase the incidence of Crown Rot and Dry Rot, while the large roots, which have a lower dry matter content are more susceptible to side splitting. Bacterial rots are then able to enter and decay of the affected roots ensues. For this reason the population should not be allowed to fall below 24,000 plants per acre. As some plants are always lost before harvest, the final 'stand' after either drilling 'to a stand' or singling should not be less than 27,000 plants per acre.

As the size of the individual roots depends to a high degree on the plant density, the optimum plant population will in turn depend on the size of swede that is required.

Small, neat, even bulbs are necessary for swedes grown for human consumption in the south of England; tops are valueless and should be minimal.

When swedes are required for folding in situ by sheep, especially in late winter and on exposed uplands, good frost hardiness is essen-tial. Relatively small bulbs with a fairly tight canopy of leaves have been observed to be much hardier than a more open crop with larger roots grown at a wider spacing.

On the other hand, when the crop is to be removed from the field for feeding, larger roots are more convenient for handling.

When stored under cover for later use a high degree of hardiness and a good yield of top are of less importance. 'Optimal' plant densities for all situations have yet to be defined by critical work.

With the introduction of seed dressings and the widespread adoption of the precision drill, the old-fashioned practice of sowing turnips and swedes thickly at seed rates of up to 3 lb or even 4 lb per acre and then singling the resultant mass of plants has largely disappeared. The cultural requirements for precision drilling are discussed in chapter 7—for seed rates, see Table 25 (Chapter 7).

It is essential to use only graded seed which has been treated with a seed dressing recommended by the Ministry of Agriculture to control flea beetle.

It is also essential to use the correct belt or wheel for the grade of seed (and vice-versa) which is now normally graded between 6/64 in by 1/64 in to 8/64, the precise limits of the grade depending on variety and season. Each grade is designated by a letter.

The growth rate of swedes has been shown to be influenced by the size of the seed. The plants from the largest seed initially make the most vigorous growth and produce the largest leaf area; plants from the smallest seed make the poorest showing.

While the germination of the various sizes within any given sample is the same, plants from the largest seed show a slightly better rate of emergence. However, as growth proceeds, the plants from the smaller seeds grow relatively faster and, by harvest, much, but by no means all, of the difference has disappeared. At harvest the crop from large seed still yields appreciably more than the smallest seed and slightly more than commercially 'graded' seed.

It is therefore important that merchants supply as high a proportion of large seeds within the graded fraction as is economically possible. Two grades might result in the sowing of the highest possible proportion of large seed but this course is likely to complicate commercial practice.

Pelleted seed, which is currently in trial, might prove to be a preferable alternative. If successful, it would allow a constant grade of pellet to be used, together with the whole of the large fraction of a given parcel of seed. Only seed too small to produce a vigorous plant need then be discarded.

When swedes are precision drilled and the unwanted plants removed, a theoretical spacing of 2 in between the seed has proved to be a very safe way of establishing an even stand; 1 lb and $\frac{3}{4}$ lb seed are required for the traditional 22-in rows and 28-in ridges respectively. Singling to approximately 9 in and 8 in on 26-in and 28-in ridges and $10\frac{1}{2}$ in on 22 in rows respectively gives 27,000 plants per acre, which is suitable for large swedes and is the minimum number of plants that should be left (see Table 7).

In practice growers drill at spacings varying from 1 to 4 inches and chop out the unwanted plants. It has been shown experimentally that providing enough plants are left some irregularity of spacing does not affect yield.

TABLE 7. Plant Population Guide

Number of plants (000 per acre) as influenced by spacing between plants and row width.

Per chain	Plants Per 100 inches	Inches apart	Plant Population in Thousands/Acre				
			Row Width in Inches				
			18	20	22	24	28
54	6·8	14·7	23·8	21·4	19·4	17·8	15·3
56	7·1	14·1	24·6	22·2	20·2	18·5	16·0
58	7·3	13·7	25·5	23·0	20·9	19·1	16·4
60	7·6	13·2	26·4	23·8	21·6	19·8	17·1
62	7·8	12·8	27·3	24·6	22·3	20·5	17·6
64	8·1	12·4	28·2	25·3	23·0	21·1	18·2
66	8·3	12·0	29·0	26·1	23·8	21·8	18·8
68	8·6	11·6	29·9	26·9	24·5	22·4	19·4
70	8·8	11·3	30·8	27·7	25·2	23·1	19·9
72	9·1	11·0	31·7	28·5	25·9	23·8	20·5
74	9·3	10·7	32·6	29·3	26·6	24·4	21·0
76	9·6	10·4	33·4	30·1	27·4	25·1	21·7
78	9·8	10·2	34·3	30·9	28·1	25·7	22·1
80	10·1	9·9	35·2	31·7	28·8	26·4	22·7
82	10·4	9·7	36·1	32·5	29·5	27·1	23·2
84	10·6	9·4	37·0	33·3	30·2	27·7	24·0
86	10·9	9·2	37·8	34·1	31·0	28·4	24·5
88	11·1	9·0	38·7	34·8	31·7	29·0	25·0
90	11·4	8·8	39·6	35·6	32·4	29·7	25·6
92	11·6	8·6	40·5	36·4	33·1	30·4	26·2
94	11·9	8·4	41·4	37·2	33·8	31·0	26·8
96	12·1	8·3	42·2	38·0	34·6	31·7	27·1
98	12·4	8·1	43·1	38·8	35·3	32·3	27·8
100	12·6	7·9	44·0	39·6	36·0	33·0	28·5
Yards per acre			9680	8712	7920	7260	6255

This table has been extracted from a more detailed table prepared by and is reproduced by courtesy of the National Institute of Agricultural Engineering, Wrest Park, Silsoe, Bedford. Plant populations on 28 in rows calculated by the author.

Under experimental conditions the yield of roots at 9 in spacing on 28-in ridges was unaffected by leaving 25 per cent gaps plus 12½ per cent doubles at singling.

Singling

Less skill is required to single spaced plants and, with precision drilling, the crop can be left for a longer time before singling without loss of yield.

When singling is delayed until the seedlings have formed a rosette of leaves a shorter hoe blade is recommended (6 in compared with the normal 6–8 in). However, while hand singling of precision drilled stands may well be worthwhile when crops of 30–40 tons per acre of large swedes are to be grown, it would appear to be quite impossible to justify the practice with higher plant populations or where lower yielding crops are produced.

When the crop is completely mechanised a close spacing may be used in conjunction with down-the-row thinners or gappers.

Alternatively the seed may be drilled to a stand. The latter method is becoming increasingly popular in Wales and in swede-growing areas in the south of England but not, as yet, in Scotland.

In Wales, where small hardy swedes touching in the row are wanted, precision drilling the seed to a stand at a spacing of 4 in between the seed in rows 22 in apart is widely recommended; $\frac{1}{2}$ lb seed is required per acre. Spatial arrangements vary considerably when swedes are grown for human consumption.

In Devon 22-in rows with a 5–6-in spacing between seeds is common practice. With improved techniques for weed control interest is increasing in rows 14–18 in wide, which appear to give less trouble from splitting and rotting of the roots.

Turnips grown for bulbs may be precision drilled or are sometimes broadcast. Little critical evidence is available but plant populations should certainly not fall below those required for swedes. Some growers plant the seed 3–4 in apart with a precision drill and chop out every other plant, sowing as little as 6 oz seed per acre. The seed may be perfectly well drilled to a stand. Autumn stubble turnips should be sown at 2–4 lb per acre. The higher rate is preferable where a thick stand of tops is required; the lower rate where larger bulbs are of prime interest.

MANGOLDS AND FODDER BEET

The available varieties of mangolds and fodder beet form a continuous series. Mangolds may be divided into two groups—

The low dry matter type (10–12 per cent: Includes the majority of the English varieties of mangold. Although these have a lower dry matter content, their flesh is softer and they are less inclined to damage the teeth of young cattle and lambs. They sit well on top of the ground and are easy to lift.

The tops of this type are much smaller than those of the higher dry matter types and are not fed to stock. Both globe and intermediate shapes are available. Skin colour may be either yellow, orange, red or white. This type of mangold gives the greatest weight of roots and yields of over 80 tons per acre have been recorded.

The medium dry matter varieties (12–15 per cent): Predominantly of continental origin, their main advantage is that they usually give a higher yield of dry matter per acre than the previous type, although the roots are somewhat smaller; they usually have larger tops.

Most varieties of fodder beet are of Continental origin, with dry matter contents ranging from 15–20 per cent. There are marked differences in appearance and growth habit. Some are comparable to mangolds, with intermediate to pear shaped roots while others are almost identical in appearance to sugar beet.

In some varieties, particularly those similar to sugar beet, the roots grow deeply into the ground and are relatively difficult to lift. They can also become fangy and consequently may be very dirty when lifted. Really dirty fodder beet are unsuited to stock feeding; washing can generate considerable difficulties with soil disposal and blocked drains. If very dirty beet is fed, death can occur by blockage of the digestive tract. The tops, especially those of the varieties similar to sugar beet, are large and may be fed to cattle. Using the tops can increase the yield of dry matter per acre by as much as 50 per cent. Yields of tops usually vary from 10–12 tons per acre. Varieties with a medium dry matter content are more suited to cattle while those with the highest dry matter are primarily for pigs.

There are considerable variations in stocks of mangold which have the same varietal name but are offered by different seedsmen. The best stocks in NIAB trials come from the following varieties—

Mangolds, Low dry matter, 10–12 per cent—**Globe Shape:** Prizewinner (Yellow skin colour). **Intermediate shape:** Brocks Red (Red); Orange Intermediate 601 (Orange); Red Intermediate (Red); White Knight (White). **Medium Dry matter** (12–15 per cent)— Pajbjerg Ideal P. (Yellowish Orange).

FODDER BEET (15–20 per cent dry matter)—Monorosa, monogerm variety (Red); Pajbjerg Korsroe P. (Yellowish Orange); Red Øtofte Polyploid (Red); Taca Trifolium (Yellowish Orange); Pajbjerg Rex P. (White); White Øtofte (White).

Sowing date

Sowing usually occurs between mid-April and mid-May. Varieties liable to bolt should not be sown too early, as this may precipitate excessive bolting, especially in the north. Orange Intermediate 601 which shows a low susceptibility to bolting, is suitable for the earlier sowings. The best results are probably obtained from sowings in the latter part of April and the first few days in May; timely sowing catches the soil moisture and allows the crop to get established before the advent of really hot weather. Delaying sowing until late May usually results in substantial reductions in the yield of dry matter.

Seed-Rate and plant population

Before the introduction of rubbed and graded seed, natural seed was sown at 8–10 lb per acre and singled. 'Natural' seed is in fact a corky fruit usually containing one to three true seeds, tending to produce bunches of plants which are difficult to separate; this, combined with the relatively high seedrate, has meant that singling of such stands is a relatively slow and time-consuming job.

Rubbed and graded seed $\frac{8}{64}-\frac{10}{64}$ in suitable for precision drills is available from most seedsmen, and should be spaced 2–3 in apart within the row. The seedrate depends on the row width and is shown in Table 25 (Chapter 7).

Pelleted seed, $\frac{9}{64}-\frac{12}{64}$ in is also available in a limited number of varieties from some seedsmen and is approximately four to five times heavier than graded seed.

In view of the importance of obtaining the largest possible yield of dry matter from a small acreage of mangolds or fodder beet, it is debatable whether the risks of drilling to a stand are worthwhile. It is often much safer to drill more plants than are needed and chop out those not required. Higher populations are necessary on soils of lower fertility. In order to obtain at least 25,000 plants per acre at harvest, it is best to aim for 28–30,000 plants at singling; the precise singling distance depends on row width (Table 7). Although yield is not affected by a limited degree of irregularity of spacing, it is essential to leave enough plants.

Boron deficiency of swedes and turnips

Boron deficiency is also known as Heart rot, brown heart, internal browning or as Raan in Scotland. The centres of affected roots become fibrous and greyish brown; these symptoms cannot be detected from the outside.

Boron deficiency occurs in soils which are deficient in boron or on soils which have been limed recently; the trouble can be induced by overliming light soils.

The incidence of this condition may be prevented by the application of boronated fertiliser but the treatment should be confined to soils where boron deficiency is known to occur. Excessive applications of boron can induce boron toxicity when the growth of swede and turnip plants becomes stunted.

The manuring of swedes and turnips

Owing to the increased danger of clubroot infection under acid conditions it is essential that swedes and turnips are only planted on soils that are not acid; acidity must always be corrected before planting.

Unlike the other brassicas, swedes and turnips do not respond to liberal applications of nitrogen, in fact the reverse is the case. Excess nitrogen, applied either as farmyard manure or as artificial fertiliser results in watery roots which are of low feeding value and of poor keeping quality. Too much nitrogen also results in excessive foliage production and large coarse necks on the swedes; such roots are not suitable for the vegetable market.

The appropriate rate of nitrogen application is shown in Table 8 and depends on the previous crop, the farming system and whether farmyard manure has been given; only moderate dressings, say 10–12 tons per acre should be applied.

TABLE 8. Recommendations for the Nitrogenous Manuring of Swedes and Turnips*

GROWN AS MAINCROPS FOR BULBS	*Units N per acre*
Permanent grass or ley grazed and in good heart.	
Usually high clover content	20
Permanent grass or ley with low clover content	40
Arable soils in good heart	
With FYM	30
Without FYM	40–50
Poorer arable soils	
With FYM	50
Without FYM	65
GROWN AS CATCH CROPS FOR BULBS/TOPS	
Heavily manured early potatoes	NIL
Ley (grazed)	20
Ley (mown for hay or silage)	40
Cereal (followed by stubble turnips)	50–60
Cereal (late sown turnips for tops only)	40–50

*Recommendations for direct drilled crops are given in the appropriate chapter, these are generally about 20 units per acre *above* those for conventionally drilled crops, although slightly more generous treatment should be given in poorer situations. The rate of application for sandy soils should be increased by up to 20 units per acre. The nitrogen application to swedes for human consumption grown in Devon is usually restricted to between 20–40 units per acre; excess produces heavy necks.

It is most convenient to apply dung before the autumn ploughing; in the north dung has traditionally been applied to the ridges but this involves a heavy labour requirement at a busy time of the year.

Farmers growing swedes for human consumption in Devon are particularly careful with their nitrogen application; some apply as little as 24 units N per acre.

Recommendations for the rate of application of phosphate and potash are shown in Table 9. Generous amounts of phosphate and potash are of the utmost importance where soil reserves are low, as on newly reclaimed sites. Although smaller applications are more economical where soil reserves are adequate, phosphate is still relatively cheap; building up soil reserves is sound policy, as this usually results in improved yields from succeeding crops.

In wet areas, phosphate may be applied either as basic slag or ground mineral phosphate (Gafsa phosphate). Stubble turnips, especially on soils well supplied with phosphate and potash require relatively small applications, 30–40 units each of P_2O_5 and K_2O are normally ample.

TABLE 9. Recommendations for the Phosphate and Potash Manuring of Swedes and Turnips

	Units per acre	
	P_2O_5	K_2O
Standard dressing:		
With FYM	70	70
Without FYM	100	100
Low soil P_2O_5* or K_2O:		
With FYM	100	100
Without FYM	125–150	125–150
Good soil reserves of P_2O_5 and K_2O:		
With FYM	40	40
Without FYM	50	50
Sown as catch crop after cereal according to soil reserves—little needed for late sown turnips for tops	0–40	0–40

*On newly reclaimed land or in high rainfall areas 10 cwt high grade basic slag may be applied which may replace up to 70 units of P_2O_5 in the recommendation. Ground mineral phosphate may also be used in areas with over 35 in of rainfall. Additional potash may be applied as 60 per cent muriate of potash if required.

The fertiliser is broadcast and worked into the seedbed. When the crop is grown on ridges, broadcasting on the flat prior to ridging results in maximum concentration in the ridges. Fertiliser should never be placed in contact with the seed.

Manuring of mangolds and fodder beet

Mangolds are an expensive crop to produce, so that it is essential to grow a really heavy weight of roots to make the crop worthwhile. Very high fertility is necessary to achieve such a crop and a really liberal application of well rotted farmyard manure should be ploughed in during the autumn; about 20 tons per acre or as much as can conveniently be turned under should be given.

A generous dressing of complete fertiliser should be worked into the soil during the course of seedbed preparation; recommended dressings are shown in Table 10.

TABLE 10. The Manuring of Mangolds and Fodder Beet

Standard dressing for all soils*:	Units per acre		
	N	P_2O_5	K_2O
With FYM	100	100	120
Without FYM	120	120	150

*Apply as compound fertiliser with ratio $1:1:1\frac{1}{2}$.

The fertiliser dressing may be reduced somewhat with very heavy dressings of manure or extremely high fertility. However, ample fertiliser must be given to ensure maximum yield.

The small acreages of these crops do not permit any very great economies to be effected by reducing the fertiliser application.

Chapter 4

THE BRASSICA GREEN FORAGE AND ROOT CROPS

CLIMATE IS the main factor which determines the yield of turnips and swedes and largely accounts for the wide variation in yield between one area and another. The most suitable areas are those with relatively cool summers, where neither extremes of drought nor wetness are experienced. With the exception of catch crop or stubble turnips, for which records are not available, these crops are now largely confined to Scotland, the north and west of England and parts of Wales.

Almost half the area of turnips and swedes recorded for England and Wales was grown in Yorkshire and in the counties further north, where yields are generally high. Although Devon grew the third largest acreage of any county in England and Wales, the crops are sown later and the yields usually compare unfavourably with those of the crops grown further north; the strip along the south Devon coast is frequently too hot and dry for the production of satisfactory crops.

The importance of turnips and swedes in Brecon and Radnor is worthy of notice; this area has the heaviest stocking rate of sheep in the United Kingdom (218 ewes per 100 acres of crops and grass at the June 4th census in 1970) and also grows the greatest proportion of swedes, 17·1 acres per hundred acres of arable land.

Only in these counties has the area of turnips and swedes grown in 1970 (6,478 acres) risen once again to approach that cultivated in 1922 (8,509 acres); this increase could well be a portent for the future if modern techniques are adopted in areas where the climate is well suited to growing heavy crops of swedes.

Turnips and swedes are far more important in Scotland than in most areas of England and Wales. With the exception of Brecon and Radnor, Cumberland and one or two other counties with high sheep populations and relatively small acreages of turnips and swedes, the proportion of the arable area occupied by these crops is much higher than that in England and Wales; 1·0 and 9·0 acres of turnips and swedes per hundred acres of arable land in England with Wales and Scotland respectively. The greatest concentration of turnips and swedes is seen in Aberdeenshire and Banffshire, 14·5 acres per hundred acres of arable land, where they are by far the most important fodder crops; rape, kale and cabbages are of little consequence in this area.

Like turnips and swedes, early sowings of rape are susceptible to mildew in hot dry weather; the crop does best where there is a good supply of soil moisture. Rape is used primarily for feeding sheep and almost half the recorded acreage is grown in Wales where it has long been used in connection with the reseeding of grassland, land reclamation and fattening lambs in the autumn. Rape has largely maintained its acreage in the north east of England, notably Northumberland, the North Riding of Yorkshire and Durham; the crop is also of some importance in Devon and Cornwall. It appears that a substantial acreage, is fairly widely grown as a catch crop after cereals.

Kale is most suitable when a well drained soil and mild autumn weather allow efficient utilisation in situ. Kale is less susceptible to the effects of drought than turnips, swedes and rape and does not suffer from mildew. Although widely grown, kale is of least importance in Scotland, Wales and the extreme north of England. The acreage tends to increase as one travels south and west; the largest recorded areas are grown in the south and south west of England. The limited acreage of cabbages is also widely dispersed but the largest areas are again grown in Devon and Cornwall. In Scotland 40·5 per cent of the recorded acreage of kale and cabbage is grown in the south-western counties.

Once established, mangolds do well in hot dry summer conditions; they are therefore far more suitable than swedes in the hotter and drier parts of the country. The crop is widely distributed but is of negligible importance in Scotland.

PLACE IN THE ROTATION

Cattle cabbage, mangolds, fodder beet and swedes grown in Scotland and the north of England are taken as maincrops in the traditional 'root break' and occupy the land for a full growing season; it is then essential to obtain really heavy yields to make these crops financially worthwhile. On the other hand, the kales, rapes, fodder radish and white turnips, which produce their full yield of digestible dry matter in a fairly short time, are best grown as catch crops; allowing a longer growing period increases the total yield of dry matter but much of this increase becomes fibrous and inedible.

Maincrops are generally grown between two corn crops and frequently receive a heavy dressing of dung; they may also follow rich ploughed grassland.

The earlier sown catch crops, such as kales, rapes, yellow turnips and sometimes swedes may be taken after a cut of silage or hay or a crop of early potatoes.

Late sown catch crops, rape, rape kales, white turnips or fodder radish are usually planted after an early harvested cereal or grassland that is broken late in the summer.

Winter cereals, which may be used to provide early spring grazing and are harvested in July or early August, are suitable to precede the catch crop.

Growing susceptible brassica crops more than about once in four years is generally inadvisable as it often results in an outbreak of clubroot. The problem may arise when frequent catch crops of rape are taken; catch cropping should therefore be carefully planned in advance and should not be conducted in a haphazard manner.

DISEASES AND PESTS

Clubroot (*Plasmodiophora brassicae Wor.*)

Clubroot is the most serious disease which affects the brassica crops. The causal organism attacks the roots on which it forms large 'clubs' or nodules. These usually prevent the uptake of water and nutrients, so that affected plants often wilt and die.

The presence of the disease may be confirmed by cutting the 'clubs' open. If the disease is present, the cut section will present a typical 'marbled' appearance. Ultimately the clubs disintegrate in a wet rot or slime, releasing their spores into the soil, where they can survive for seven years or longer without recourse to another susceptible 'host' plant.

Once affected, a field must be rested from susceptible brassicas for a long period. Where intensive catch cropping is practised, mixing rape or turnips with Italian ryegrass is probably best avoided. Italian ryegrass by itself gives at least some break from brassica crops. Clubroot can also be carried by cruciferous weeds such as runch, charlock or shepherds purse.

While overcropping with brassicas should be assiduously avoided, frequency of cropping is only part of the story.

Prevention is the only possible course. Other means include

drainage, liming, general sanitation, chemical means (transplants only) and the growing of resistant types and varieties.

Like many other parasites the clubroot organism thrives under certain specific conditions. By making the field conditions unsuitable for the development of the disease, its spread can often be prevented or at least greatly reduced. The resting spores germinate and form motile spores which swim about in the soil solution and infect the root hairs of the young plants, which they enter readily. The disease is therefore worst in the wetter areas of the country, notably the north and west, especially so when the drainage of the soil is impeded. It is therefore essential to see that fields growing brassicas are well drained.

Clubroot also thrives under acid conditions. For this reason alone, brassicas should never be grown on acid soils, these should always be limed first and brought up to a pH of 6·5 or more, according to soil type. Several successive crops of susceptible brassicas have been grown without trouble on well drained sites which were limed frequently, although the practice cannot be recommended with assurance. The safest course is not to grow susceptible brassicas too frequently and also to maintain a satisfactory soil pH.

General sanitation is necessary to keep the disease from infecting clean fields; the measures taken cannot be too thorough, as the disease is highly infectious. When transplanting, as with cabbages, it is essential to ensure that the plant bed occupies clean soil; plants are most susceptible in their young stages. There is no more effective way of inoculating a field with clubroot than by planting infected brassica plants at regularly spaced intervals. The plant bed should also be on land free from other pests such as potato root eelworm, as these can equally well be carried on soil attached to the roots and be ready to attack their host crops in due course.

As the clubroot organism is not destroyed by passing through an animal's gut, it follows that the dung of animals fed on affected swedes and turnips is likely to be highly infective. The best course is to feed infected crops in situ and not allow the stock access to any ground other than permanent pasture which will not be ploughed, until the stock have had time to void all the spores; a period of seven to ten days is probably desirable.

Clubroot spores may also be transported in dirt sticking to the sheep's feet. It might also be helpful to run the sheep through a formalin footbath before they are moved to clean ground; in any case this is a normal technique of animal husbandry.

On no account should infected roots be fed on land likely to be planted with brassicas. When infected roots are fed on pasture and the land is then cropped with susceptible types or varieties, clubroot invariably follows; pastures which have not been previously ploughed for 30 or more years have been known to become badly affected and the first crop of swedes taken after ploughing has been ravaged by clubroot. Similarly, clubbed swedes or turnips should not be fed to stock in buildings and the dung then be carted out on to arable land.

Chemical prevention is only practicable with transplants such as cabbages, brussels sprouts and cauliflowers. Pure calomel at 1–2 oz per pint is made into a slurry which the roots are dipped immediately prior to planting.

The use of clubroot resistant types and varieties provides a very promising approach to the problem.

TABLE 11. Types and Varieties of Brassicas which Show Good Resistance or Tolerance to the Common Races of Clubroot

Crop	Type or variety	Remarks
B. oleracea kales	Marrow stem kale Thousand head kale	Only susceptible to common races under heavy infestations and very acid conditions.
	Maris kestrel	Appears to be slightly more affected than marrow stem and thousand head kales.
Rape	Nevin	Clubroot tolerant roots may become heavily clubbed but tops do not collapse. Lower yield than other types and only suitable for situations where clubroot is present.
Fodder radish	All varieties	Can be affected but rarely damaged.
Turnips	*Continental stubble turnips:* Barive, Debra, Gelria R, Labra, Novitas, Tigra, Vobra	These varieties show the highest degree of resistance to the normal races of clubroot as stated in the Netherlands 'rassenlijst 1972'.
	Yellow turnips: The Bruce (most resistant) The Wallace	Stocks of seed may vary considerably in their degree of resistance to the common races of clubroot.
Swedes	*Purple top:* Wye, Chignecto *Green top:* Danila, Resfingtoe, Wilhemsburger Øtofte, Wilhemsburger	Stocks of seed may vary considerably in their degree of resistance to the common races of clubroot. The resistance of these varieties can break down in the face of heavy infection.

The resistance of the types and varieties listed in Table 11 cannot be regarded as absolute. The degree of the resistance depends on the races of the disease organism present, the level of the soil infection and the soil conditions, especially the state of the drainage and the soil pH; even very resistant varieties can succumb when exposed to a heavy infection (Plate 9), especially on acid soils, poorly drained soils and soils situated in areas of high rainfall.

The degree of resistance of any variety may be reduced by reselection of seed stocks in the course of seed multiplication; this is known to have happened to some stocks of the Bruce and the Wallace and can happen only too readily in other cases.

It is therefore still essential that when growing 'resistant' varieties all possible precautions against clubroot should be taken, especially the application of adequate lime to the soil.

Powdery mildew (Erysiphe polygoni DC)

Powdery mildew attacks swedes, turnips and rape, producing a white powdery covering on the leaf surface. Affected leaves drop off prematurely and the yield is considerably reduced. Powdery mildew is more prevalent in the south of England, especially in hot dry summers; the disease is most serious in early sown crops, particularly those that get away to a good start.

There is no cure and the disease is difficult to prevent in hot dry weather. The most effective preventive measures include avoiding very early sowing, giving adequate nitrogen (Plate 9) and sowing varieties which are not unduly susceptible.

The following varieties show the highest degree of resistance but these can succumb to a heavy attack:

Rape: Emerald, Lair, Blako, Gartons Late Dwarf.

Swede: Chignecto, Bangholm Sahna, Bangholm, Bangholm Wilby, Ruta Øtofte, Vogesa, Wilhelmsburger.

Dry rot of swedes (Phoma lingam)

Dry rot of swedes produces conspicuous and characteristic brownish grey lesions on the sides of the bulbs, particularly in the late autumn. In due course the diseased tissue dries up, causing cracks in the bulbs through which bacterial soft rots may enter; the plants then rot and collapse.

The fungus first develops on the seedling which may be heavily damaged or even killed. Seed borne infection is a frequent source of the disease and only swede seed tested for the presence of the dry rot fungus should be planted. The best seedsmen build up their stocks from seed which has been given hot water treatment or soaked in warm fungicide.

Infection may also be contracted from brassica waste or farmyard manure and may be carried in the soil. Good sanitation is again of considerable importance. Infected crops should be fed in situ and not carted to land which may later be planted with brassicas. The remains of brassicas and uneaten roots should be destroyed or buried

PLATE 4

The effect of sowing date on relative growth of rapes and kales: (1) hybrid kale and rape.
Photographs 19th November, 1971.

Drilled on 12th July, 1971. (NIAB trials). Left: hybrid kale (Maris Kestrel)
Right: rape (Emerald). ▼

Drilled on 27th August, 1971.
Left: hybrid kale (Maris Kestrel). Right: rape (Emerald).

PLATE 5
Rape

The effect of variety, nitrogen level and sowing date. Photographs 19th November, 1971.

► Left: Giant rape (Emerald). Right: Dwarf rape (Broad leaved Essex). Supplied by NIAB from their trials at Seale-Hayne Agricultural College precision drilled on 12th July, 1971 at 2in spacing in rows 20in apart.

◄ Emerald rape direct drilled into winter barley stubble 11th August, 1971; 60 units each of P_2O_5 and K_2O applied. Left: no N applied. Right: 100lb N applied per acre.

► Emerald rape. Left: drilled 12th July, 1971 (NIAB trials). Right: drilled 27th August, 1971.

PLATE 6

The effect of sowing date on relative growth of rapes and kales: (2) dwarf thousand head kale and Hungry Gap "kale". Photographs 19th November, 1971.

Drilled on 2nd June, 1971. Left: dwarf thousand head kale (Canson). Right: rape (Hungry Gap "kale").

Drilled on 27th August, 1971. Left: dwarf thousand head kale (Canson). Right: rape (Hungry Gap "kale").

Devon flatpoll cattle cabbage.

PLATE 7
Fodder Radish

The effect of variety.
Photographs 19th November, 1971

◀

The flowers and seed pods typical of early flowering annual varieties of fodder radish currently available. The crop should be eaten *before* the onset of flowering as the stem becomes very woody and un-palatable by the time the stage in the illustration is reached. Note the restricted develop-ment of the "root".

▶

The biennial habit of the variety "Hailstone" (turnip rooted fodder radish) when sown in late summer. Note the absence of flowering stems and the highly developed turnip-like root, which may reach the size of a well grown turnip bulb.

and not be thrown on dung heaps or dumped near fields where brassicas are likely to be grown.

Bacterial soft rots

Bacterial soft rots, which may assume serious proportions in swedes and turnips, usually enter through parts of the plant damaged by frost, pests, diseases and cracks. Large roots or stems are more likely to split and low plant densities should thus generally be avoided. Varieties of swede and turnip with a low dry matter content and crops which have been given excessive applications of nitrogen, are also more susceptible to bacterial soft rots.

PESTS

Flea beetles

Flea beetles eat numerous small pits in the cotyledons of brassica seedlings. The beetles usually attack the seedlings just after emergence but can also attack before the seedlings emerge. Once the plants have reached the rough leaf stage they are much less susceptible. The beetles are most active in hot, dry or thundery weather, especially in May and can ruin a promising crop within a matter of hours.

Once a crop has been severely damaged, it rarely makes a satisfactory recovery and redrilling is usually necessary. Before the introduction of suitable insecticides kale crops often had to be drilled three or four times as the writer well remembers. Seed dressings are highly effective.

All brassica seeds should be dressed with an appropriate insecticidal/fungicidal seed dressing such as gamma BHC+Thiram as a routine measure.

Under normal conditions seed dressings protect the seedlings for up to a week after emergence. However, it is essential for the grower to watch for damage as soon as emergence starts and to be ready to spray immediately the first pits appear on the cotyledons; trouble should be anticipated in hot weather if large numbers of flea beetles are active and a precautionary spray with BHC may then be worthwhile a few days after emergence. Dusts should preferably be applied in the early morning before the dew rises; the dust then sticks well to the cotyledons.

Other pests

Other pests, notably cabbage caterpillar, cabbage aphis and cabbage root fly may attack brassica crops but apart from cattle cabbage, control measures are only economically justifiable in the case of a really heavy infestation. Cattle cabbage are worth treating at lower levels of attack and are also subject to severe damage by

E

cabbage root fly, which lays its eggs in the soil at the base of the plant; the grubs hatch out and eat the lateral rootlets and then penetrate into the taproot.

In field brassicas the wounds on the roots permit the entry of soil organisms, resulting in the occurrence of rotting. Control measures appropriate to the various insect pests of brassicas are summarised in Table 12.

TABLE 12. Control of Pests in Brassica Crops

Pest	Chemical to use	Minimum period between application and feeding	Method of application
CABBAGE APHIDS	Dimethoate or most of the common organo phosphorus systemic insecticides	3 weeks	Foliar Spray
CABBAGE CATERPILLARS	Azinphos—methyl	3 weeks	Foliar Spray
	Mevinphos	3 days	Foliar Spray
	Trichlorphon	2 days	Foliar Spray
CABBAGE ROOT FLY (Flatpoll cabbages only)	Chlorfenvinphos	3 weeks	Chlorfenvinphos and diazinon given as individual plant treatment with special soil drench or granule applicator
	Diazinon	2 weeks	
FLEA BEETLE	BHC+fungicide	—	Seed dressing
	BHC	—	Foliar Spray or Dust

N.B. Manufacturers' instruction leaflets should be consulted for dosage rates, methods of application and safety precautions.

Diseases and pests of mangolds and fodder beet

Virus yellows

Virus yellows cause the outer leaves of the plants to turn yellow and become gorged with starch, crackling when squeezed in the hand; healthy leaves or leaves discoloured for other reasons do not. The earlier this disease affects the crop the greater is the reduction in yield. Virus yellows are most likely to prove troublesome in areas with substantial acreages of sugar beet.

Prevention depends on controlling the aphid vector of the causal virus and sources of overwintering virus should be eliminated wherever possible. Mangold clamps should be cleared by early May and all litter burned or buried; mangolds should not be clamped in the proximity of fields destined to grow sugar beet or mangolds the following year. Crops should also be sprayed with a systemic organo-phosphorus insecticide (Table 13) and grown as far apart from other farmers' crops of sugar beet and mangolds as possible.

TABLE 13. Control of Pests in Mangolds and Fodder Beet

Pest	Chemical to use	Method of application
APHID (PEACH/POTATO)	Dimethoate or most of the common organo phosphorous systemic insecticides	Foliar Spray
MANGOLD FLEA BEETLE	DDT	Foliar Spray or Dust
LEAF MINER (Mangold fly)	Dimethoate formo-thion phosphamidon trichlorphon	Foliar Spray

Mangold flea beetle

Like the turnip flea beetle, the mangold flea beetle can ruin a useful stand of young seedlings very quickly, causing similar injury. The grower should therefore be on the lookout for signs of damage while the crop is in the cotyledon stage; DDT, applied as a spray or dust as soon as damage is observed normally prevents further loss. Seed dressings used to control flea beetle in brassicas are phytotoxic to mangolds and fodder beet and cannot be used. The seed is normally treated by the merchant with an organo-mercurial fungicidal seed dressing against blackleg but this treatment gives no protection against any insect pest.

Leaf miners (mangold fly)

Mangold flies lay their small white eggs on the undersides of the leaves of young mangold or beet plants, where they may be readily seen on inspection. After hatching, the maggots bore into the leaf and eat the soft fleshy tissue between the upper and lower skins; their tunnels are apparent from the surface as thick tortuous lines, which widen to form blisters. The importance of the pest depends on the weight of the attack, the size the plants have attained in relation to singling and on the growing conditions. With a light attack or with moist soil and good growing conditions, the crop usually outgrows the danger. However, with a heavy attack, a backward or early singled crop and poor growing conditions, foliar treatment with an appropriate systemic insecticide is essential.

VARIETIES AND STOCKS OF SEED

The process of selection during seed multiplication can alter the detailed characteristics of a stock of seed of roots or brassicas in only a few generations. Consequently, stocks of seed which are of the same named variety but supplied by different seedsmen may behave quite differently when planted. These differences may be so large, at least in some respects, that for all practical purposes, the two stocks may be considered to be two different varieties.

Similarly, in the hands of a single seedsman, the characteristics of a given stock of seed may be altered profoundly during a period of only three or four years. It is therefore by far the most satisfactory course to buy seed from a seedsman who has a good stock of the required variety.

The selection should be made from the series of Farmers' leaflets issued by the National Institute of Agricultural Botany, Huntingdon Road, Cambridge, which carries out a continuous series of variety trials at a number of centres throughout the country. The relative yields and characteristics of the varieties and seed stocks which give the best performance are shown; the leaflets are revised regularly.

Wherever possible authentic seed of the root and brassica forage crops should be obtained direct from the breeder or his accredited agent, as certified seed of these crops is not generally available. Smaller merchants usually obtain their stocks from the larger whole-sale houses and breeders; the original source of the seed should then be checked.

British merchants are usually able to supply seed of foreign varieties. When certified seed is available, which is the normal case with the grasses, it is advisable to buy it. Seed certification en-sures that not only is the parcel of seed offered free from disease and the seeds of weeds and of other species and varieties, but that it is of the stated variety and true to type as originally produced by the breeder. Furthermore, the source of seed is of importance in relation to seed borne diseases such as dry rot in swedes; wherever possible seed should be obtained from national seed merchants of high repute.

FEEDING VALUE

The roots and brassica green forage crops are a very useful source of cheap energy for cattle and sheep. Provided they are fed at an appropriate stage of growth these crops are highly digestible. They also maintain their digestibility over a relatively long period, although the duration of this period depends to a large extent on the crop and on the variety.

At one extreme, early flowering varieties of fodder radish behave rather similarly to grass and only combine a worthwhile yield with a high digestibility for a very short period—probably a week or two, depending on the weather. At the other extreme mangolds and fodder beet maintain a high digestibility until well into the spring; stock relish them even when there is ample grass in May.

Although the dry matter content of the root crops is usually low, varying from 7–12 per cent in turnips, swedes and low dry matter mangolds and from 13–20 per cent in medium dry matter mangolds and fodder beet, the dry matter is rich in simple sugars. Roots are therefore exceedingly palatable and are among the most digestible foods available for winter feeding. The relatively low dry matter content of these crops is thus offset to a considerable degree by the high concentration of readily available nutrients in their dry matter.

Storage has been shown to increase the sugar content of swedes, which are a particularly suitable food for late winter and early spring. Such is also the case with mangolds and fodder beet, although the latter have considerably higher dry matter and sugar contents. Roots generally have a low crude protein content which is only about half that of marrow stem kale.

While heavy reliance on roots tends to reduce dry matter intake and consequently the performance of the stock, feeding moderate quantities of roots as part of a properly balanced ration has the reverse effect. Limited quantities of roots can thus increase dry matter intake, improve performance and reduce feeding costs.

In trials at the Rowett Research Institute, steers fed on a mainly barley grain diet gained 2·4 lb per day and required only 5·9 lb dry matter/lb liveweight gain but steers fed on swedes alone gained 1·8 lb per day and required 7·1 lb dry matter/lb liveweight gain.

Steers receiving 33 per cent of their dry matter intake as swedes and 64 per cent as barley grew as fast as those receiving only barley and showed a similar efficiency of feed conversion. All stock in this trial were bedded on barley straw. Other workers obtained the highest dry matter intake with sheep when they fed 33 per cent of the dry matter as swedes.

Experiments in New Zealand have shown that when half the daily intake of dry matter of dairy cows at grass was replaced by either silage or turnips, the cows receiving grass and turnips produced 9 per cent more milk and butterfat than those receiving grass and silage. Similarly, when turnips or fodder beet have been offered in conjunction with a diet of maize silage, the butterfat content of the milk is increased.

With the exception of flat poll cabbage, the leafy brassica forage crops usually have a somewhat higher dry matter content than turnips, swedes and the traditional type of mangold but their digestibility is more variable; the digestibility of the stems usually declines rapidly as fibrous tissue is laid down. Digestibility of leaf and stem may thus differ enormously, especially in the thousand head kales, where a high proportion of the stem becomes very fibrous and is then totally neglected by stock (Table 14). The high digestibility of the stem of Maris Kestrel should be noted; the pith of this variety is also richer in soluble dry matter (sugars etc.) than the pith in the stems of ordinary marrow stem kale.

TABLE 14. Relative Digestibility of Leaf and Stem in Kale
Digestible dry matter content %.
NIAB main thousand head kale trials at Cambridge; January, 1968

	Leaf	Stem
Normal thousand head kale		
(average of 12 varieties)	80·8	62·9
(range of 12 varieties)		
highest reading	82·8	66·6
lowest reading	78·8	58·8
Dwarf thousand head kale		
(average of 3 varieties)	81·8	70·3
(range of 3 varieties)		
highest reading	82·0	72·2
lowest reading	81·6	67·4
Maris Kestrel	82·4	82·8
Proteor	80·6	72·4
Marrow stem (Carters)	80·4	75·0

Published by kind permission of the National Institute of Agricultural Botany, Cambridge. These are a single year's results and are given for illustrative purposes only.

Brussels sprouts are currently being crossed with marrow stem kale hybrids to increase the dry matter and soluble dry matter contents of the stems of kale.

Freshly grown crops of rape are also highly digestible; if the crop is allowed to become mature, the digestibility of the stems decreases considerably. Rape has a high crude protein content of around 23 per cent of the dry matter.

In practice the digestibility of these crops depends on the parts of the plant actually eaten by the stock. If stock are to obtain a highly digestible diet, the crop should be in an appropriate stage of growth for folding and the animals should not be penned too tightly otherwise their dung becomes hard, indicating constipation. When sheep fed on protein rich leafy young brassicas, receive concentrate supplementation, it is quite sufficient to provide rolled cereals and minerals. Extra protein is best omitted.

Dangers

The roots or bulbs of turnips and swedes are normally remarkably safe foods and may be fed in quantity to fattening cattle and sheep without toxic effects. Heavy feeding of turnips or swedes to dairy cows should be avoided as it is likely to taint milk; they should be given immediately after milking.

Growing or newly harvested mangolds contain a small amount of nitrate and possibly a number of other toxic substances; if fed to animals at this stage, they are likely to cause illness or sometimes death. Consequently mangolds are allowed to ripen in a clamp and not fed to stock until after Christmas.

Mangolds grown on lime-deficient soils constitute a much greater risk, as the oxalic acid and or oxalates, which are normally present in greater quantity than in other crops, are only partially reduced during storage. If such roots are fed in quantity or over a long period, poisoning of the stock occurs.

Provided the roots are grown in a soil which is adequately supplied with lime and they are properly ripened, mangolds are a perfectly safe feed in reasonable quantities.

Sugar beet and fodder beet tops also contain quantities of oxalic acid and or oxalates and should be properly wilted before feeding. Sometimes they are left so long that only the crowns remain but these are relatively innocuous. Oxalate poisoning may be prevented by feeding chalk. It is inadvisable to feed frosted roots and can prove to be dangerous.

Metabolic disorders

The leafy brassica forage crops, including brussels sprouts, contain substances which can cause severe metabolic disorders and even death of the stock if they constitute a high proportion of the ration, or stock are allowed to gorge on them. These substances are concentrated in the leaves of the crop; swede leaves have also proved to have toxic properties when fed in large quantities. Individual crops vary in the degree to which they may prove toxic to stock from field to field, from one area to another and from season to season. Both sheep and cattle may be affected.

Haemolytic anaemia can occur with varying degrees of severity in cattle and sheep, although the latter are much less susceptible. The first visible symptom, generally apparent from 10–18 days after the start of heavy kale feeding, is usually but not invariably bloodstained urine (haemoglobinurea), which foams as it falls to the ground; clinical symptoms of anaemia follow, including weakness and loss of appetite.

The only way of preventing the occurrence of haemolytic anaemia is to limit the intake of kale or other leafy brassicas by the stock to a maximum of 56–60 lb per head per day for cows of the larger breeds like Friesians and 40–45 lb per day for cows of the smaller breeds such as Jerseys and Guernseys.

If the symptoms of haemolytic anaemia are seen, the feeding of all leafy brassicas should be stopped immediately; recovery is then usually rapid. If the animals are allowed to continue eating leafy brassicas, deaths are likely to occur.

Mineral deficiencies

It has been known for many years that brassicas contain goitrogenic substances which limit the uptake of iodine by the thyroid gland, resulting in an 'induced' iodine deficiency in the stock; the feeding of leafy brassicas doubles their iodine requirement. Iodine deficiency is likely to occur when the diet of farm animals contains a large proportion of leafy brassicas; stock fed on roots appear to be unaffected.

Sheep and cattle in late pregnancy and high yielding cows in the earlier part of lactation are at greatest risk. Pregnancy may be terminated by abortion; if the gestation period runs the full term the young may be either still-born or die shortly after birth, a normal symptom being an enlarged thyroid gland in the foetus.

Iodine deficiency may also cause serious infertility problems in dairy herds; this deficiency may also prove to be a problem on farms in some high rainfall areas, where the heavy feeding of leafy brassicas may aggravate a difficult position still further. Grass feeding does not help as this too can be low in iodine; the problem is likely to be worst on intensively managed pastures which are liberally dressed with nitrogen and potash.

Fortunately, the effects of the type of goitrogen (thiocyanate) normally found in the leafy brassicas can be counteracted by iodine supplementation. Lambs fattened on kale or rape may also suffer from iodine deficiency but the condition is relatively mild and does not appear to affect their rate of growth; the disorder may again be prevented by feeding iodine. Ewes fed on leafy brassicas prior to lambing must receive iodine supplementation; similarly, it is frequently advisable to feed additional iodine to dairy cows consuming brassicas in the latter part of pregnancy and until they are in calf again, especially in iodine deficient areas.

Infertility problems may also be caused by other mineral imbalances and deficiencies. Leafy brassicas have a very high calcium: phosphorus ratio, which is especially wide in the leaves (Table 15): this imbalance should be corrected by feeding a readily available high phosphorus supplement.

TABLE 15. Mineral Content of Brassica Crops
% DM

	Na	K	Ca	Mg	P
Swedes					
Roots	0·14	2·35	0·45	0·08	0·20
Tops	0·28	2·73	1·96	0·17	0·17
Kale					
Leaf	0·32	2·45	2·51	0·13	0·25
Stem	0·32	3·41	0·88	0·16	0·30

(after Kay M. 'Feeding value of brassica fodder crops' paper No. 8. Conference on 'The future of brassica fodder crops'. 11.6.71 at Rowett Res. Inst. pp. 49–55).

Other foods with a high calcium content such as legume rich hay and silage or sugar beet pulp should be omitted from a ration containing kale. Where economically feasible, kale should be balanced by the inclusion of foods with a relatively high phosphate content such as the cereal grains and groundnut cake. Kale also tends to be deficient in manganese and its high calcium content may reduce the availability of the already limited supply. On farms, where there is a tendency to manganese deficiency, as on the chalkland areas, the feeding of kale may accentuate infertility problems and supplementation with manganese is then usually advisable.

Infertility problems

Infertility is a very complex problem and may occur for a number of reasons, such as disease or a low plane of nutrition; unsatisfactory mineral nutrition is only one of the possible causes. When infertility problems arise it is thus essential to call in a veterinary surgeon and to adopt treatment on his advice. 'Hit or miss' mineral supplementation is unlikely to provide useful results; if qualified professional advice is not obtained the problem could well become much more acute.

Rape poisoning

Most of the rape grown in the British Isles is fed to sheep, which are rarely adversely affected. Rape poisoning in cattle is infrequent but a number of animals may be affected simultaneously on the same farm. There are several types of rape poisoning. The type known as haemoglobinurea is the least important; serious trouble may be prevented by removing the stock as soon as symptoms occur. Other forms, which may affect the respiratory, alimentary and nervous systems severely, are often fatal and prevention is the only satisfactory course.

Rape poisoning is most likely in wet seasons and after early frost, especially on low lying or poorly drained ground. Rape showing purple discoloration is said to be particularly dangerous.

To prevent rape poisoning, stock should be introduced gradually and only allowed to graze the crop for a short time at first. The grazing time should be slowly increased until no further restriction is necessary. The stock should not be introduced to the crop in wet or frosty conditions and should receive a feed of hay first. Palatable roughage should be available at all times. Access to a permanent pasture or older ley and the admixture of turnips with the rape seed are also valuable for reducing the intake of rape. When lambs or hoggetts are to be fattened on rape they should first receive preventive treatment for pulpy kidney disease.

Nitrate poisoning

Root and brassica green fodder crops have also been reported to produce nitrate poisoning in both cattle and sheep. The trouble is most likely when these crops are grown on soils with a very high nitrogen content. The level of fertiliser nitrogen should therefore be controlled to suit the soil conditions and sowing date. Late top dressing has been known to produce toxic effects and should be avoided.

Other preventive measures include restricting the intake of leafy brassica crops and sometimes of folded swedes as well by allowing a runback on to grass and providing adequate palatable roughage. In general, crops which have been checked by drought and then grow rapidly after rain can prove to be particularly toxic.

PLANNED RATION

A properly planned ration is essential to obtain the best results from any food or system of feeding; the brassica green forage crops and the root crops are no exception to this rule. While the leafy brassica forage crops can produce severe metabolic disorders if relied on too extensively, they are most unlikely to do so if fed as an integral part of a balanced ration which includes appropriate mineral supplementation.

Similarly, although the roots of turnips and swedes are usually remarkably safe foods if fed in quantity, the 'traditional' rates of $\frac{3}{4}$–1 cwt per bullock per day are frequently wasteful; the most economic ration is more likely to lie between approximately 35 and 50 lb of swedes per head per day according to size of bullock. The following rations which include self-fed whole swedes have been used to obtain a liveweight gain of 2 lb per head per day:

	Liveweight of beast	
	6 cwt	*8 cwt*
Whole swedes	35 lb	40 lb
Hay	8 lb	10 lb
Barley	7 lb	8 lb

The value of a ration essentially depends on the quality of its constituent parts. As many foods, particularly hay and silage, are extremely variable in their composition, the only really satisfactory method of ensuring that the ration is suitable for the purpose is to have the foods analysed by a nutrition chemist; any reasonable fee involved can save a good deal of money and is usually an excellent investment.

ESTIMATION OF CROP YIELD

The estimation of fresh crop yield presents little difficulty and may be undertaken as follows.

Drilled crops

Cut a straight piece of 1 x 1½ in deal or similar light strong board, cane or iron rod to the length shown in Table 16.

TABLE 16. Length of Row Necessary to Obtain a Sample of One Square Yard from Drilled Crops of Various Row Widths
(All measurements in inches)

Row width		length of sample board	Row width		length of sample board	Row width		length of sample board
7	=	186	18	=	72	25	=	52
7	=	2 × 93	19	=	68	26	=	50
12	=	108	20	=	65	27	=	48
13	=	100	21	=	62	28	=	47
14	=	93	22	=	59	30	=	43
15	=	87	23	=	57	36	=	36
16	=	81	24	=	54			

Board or cane is much easier to use in a crop than string or a tape is more accurate; the estimation can also be carried out single handed.

Lay the board or cane against the length of row chosen for sampling and cut or pull the produce, clean any dirt off the root crops, weight and record; collect the requisite number of samples across the field, making sure that they are taken at random and are representative of the crop as a whole; they should then be averaged. The average weight per square yard may then be read against Table 17, which will give the approximate fresh yield per acre.

TABLE 17. Ready Reckoner to Convert lb per Square Yard of Crop Sample to Yield in Ton per Acre

Sample (lb)		Yield ton/acre	Sample (lb)		Yield ton/acre	Sample (lb)		Yield ton/acre	Sample (lb)		Yield ton/acre
1	=	2·0	11	=	24·0	21	=	46·0	31	=	67·0
2	=	4·5	12	=	26·0	22	=	47·5	32	=	69·0
3	=	6·5	13	=	28·0	23	=	49·5	33	=	71·5
4	=	8·5	14	=	30·0	24	=	52·0	34	=	73·5
5	=	11·0	15	=	32·5	25	=	54·0	35	=	75·5
6	=	13·0	16	=	34·5	26	=	56·0	36	=	78·0
7	=	15·0	17	=	37·0	27	=	58·5	37	=	80·0
8	=	17·5	18	=	39·0	28	=	60·5	38	=	82·0
9	=	19·5	19	=	41·0	29	=	62·5	39	=	84·0
10	=	21·5	20	=	43·0	30	=	65·0	40	=	86·5

For half pounds in sample, add one ton per acre to yield given on the above table, e.g. sample of 10½ lb per square yard = 22·5 ton per acre.
Precise yield of each lb per square yard = 2·16 ton per acre.

Small samples can show great variation and the greater the number of samples, within reason, the more accurate is the result likely to be; however farmers' time is valuable and a limited number, usually six to ten samples, is likely to give an accurate enough estimation on which to ration the crop.

In general, the more even the crop the more reliable is the estimation likely to be and the smaller will be the number of samples that it is necessary to take. It is most important that the scales should be accurate; spring balances are notorious for their inaccuracy and should be checked against government stamped metal weights before use.

Broadcast crops

The estimation of yield in broadcast crops is a little more difficult than in row crops and it is necessary to use a quadrat with an internal area of one square yard. The quadrat is best made of light angle iron and consists of a square 36 in by 36 in left open on one side. The two ends of the ⊔ so formed should have slots or loops through which a length of iron may be slid after inserting the quadrat into the crop; this closes the square.

All the plants so enclosed are cut or pulled and the samples are then weighed and calculated as for a drilled crop. An iron or steel quadrat is preferable to wood as it is much more durable and does not break at the corners; the quadrat can readily be made up by any blacksmith.

METHOD OF UTILISATION

Grazing is a cheap and effective method of feeding autumn and winter fodder crops but climate, soil and management must be right. The practice is most suited to the parts of the country which normally experience a relatively mild winter; in the more severe winters experienced in northern areas green crops are particularly liable to severe frost damage and at least part of the root crop should be stored. A spell of really hard frost can stop folding completely almost anywhere and it is essential to have an alternative supply of stored fodder.

The soil must be free draining and not subject to surface waterlogging in wet weather. Light to medium soils overlying a permeable subsoil such as chalk, limestone or sandstone are ideal; sites with impeded drainage are worse than useless and should be avoided. The elimination of soil disturbance by direct drilling reduces poaching and widens the scope of utilising roots and green crops in situ, especially with cattle.

Careful forward planning is of the utmost importance when folding root and green forage crops in autumn and winter. The field should be appropriately sited and supplied with an adequate area of run-back, preferably in the same field, so that the stock, especially ewes with lambs, do not have to pass through muddy gateways; stubble or grassland ready for the plough are ideal. Such provision prevents crowding in the fold when starting a new area, gives the stock drier lying conditions and reduces the soiling of udders in wet weather.

The run-back should include a water trough, except, perhaps, when sheep are eating large quantities of roots or other succulent fodder.

On the larger farms the fields are best divided into blocks, in which fodder crops, grass and cereals are rotated; the requisite facilities are then readily available.

The acreage of roots or green forage crop required by any given class of stock varies greatly from one situation and locality to another and is governed by the quantity consumed by the stock, the weight of the crop and the amount of waste.

The intake of the stock is influenced to a large extent by the quantity and quality of other foods on offer and their palatability relative to that of the fodder crop. The amount and proportion of waste can be extremely variable and are dependent on the maturity and palatability of the crop, the weight of the crop, the conditions of soil and weather conditions and the area of the fold. Young succulent crops tend to be eaten readily and there is usually little waste but with mature crops which have fibrous inedible stems, such as thousand head kale in the late winter, the proportion of waste may be very heavy.

The amount and the proportion of waste tend to increase with the weight of the crop; heavy crops are frequently very wasteful, especially in wet weather. Where heavy crops of swedes are to be eaten in situ, the weight of roots on offer to sheep may be reduced by carting off strips of the crop for storage or feeding elsewhere.

Conversely, the waste tends to be much lower with lighter crops as the crop is spread over a much larger area; catch crops, which tend to produce a lower yield than maincrops are particularly advantageous in this respect and much more economical.

Weather and soil conditions exert a major influence on the amount of waste, which can become extremely heavy in wet situations or in prolonged spells of wet weather; the waste associated with heavy crops then becomes so heavy as to be quite unacceptable.

The basis of sound grazing management is to provide a continuous supply of fresh succulent fodder of high digestibility—a matter of vital importance with milking and fattening stock. It is essential that the area on offer at any one time is controlled by folding or strip grazing.

Some farmers also ration their crop by imposing a time limit on the stock, which has the advantage of reducing the frequency with which it is necessary to move the fence.

In one experiment, grazing rape without restricting the amount on offer wasted 44 per cent of the dry matter and 43 per cent of the starch equivalent. When the stock are not controlled or the area of the fold is too large in relation to the number of stock, the animals walk through the crop, picking out the youngest and most succulent portions and soiling the remainder. Conversely, it is most unwise to pen stock very tightly in order to force them to clear up; such a course merely results in reduced productivity of the stock. Provided that the waste is not excessive, it is sound husbandry to regard the uneaten residue as useful green manure.

The great variation in crop yield and in the quantities of other foods that make up the ration and the variability of the amount and proportion of waste renders budgeting for the acreage required a somewhat difficult exercise; static rule of thumb recommendations are unsatisfactory as they make no allowance for these variables.

Tables 18 and 19 have been compiled for forward planning purposes and show the approximate area of fodder crop required for folding cattle and sheep at different levels of crop yield and rates of feeding. The assumed rate of wastage at 20–25 per cent is somewhat arbitrary but should normally provide adequate allowance for most practical situations; higher rates of wastage are likely with very heavy or mature crops and adverse conditions. It is usually advisable err on the generous side when making provision for the stock.

TABLE 18. Estimated Area of Kale or Stubble Catch Crop required to Provide 56 lb and 28 lb per Head per Day to Cattle for a Three-month Period, Assuming that Approximately Three quarters of the Crop is Utilised

			Area of crop required per cow/heifer/for a 3-month period	
Type of Crop	*Gross Yield (ton/acre)*	*Assumed utilised yield (ton/acre)*	*56 lb/day eaten*	*28 lb/day eaten*
Stubble catch crop poor	5·0	4·0	0·60	0·30
Stubble catch crop average	7·5	5·5	0·40	0·20
Stubble catch crop good	10·0	7·5	0·30	0·15
Kale average late sown	15·0	11·5	0·20	0·10
Kale average early sown	20·0	15·0	0·15	0·08
Kale good	25·0	19·0	0·12	0·06
Kale very good	30·0	22·5	0·10	0·05

The precise details of management depend upon the class of stock. With cattle the initial area of the fold should be large enough to avoid overcrowding of the animals. If this occurs, cattle are very likely to break through the electric fence and may then become habitual 'fence breakers'. The shape of the feeding area should be

TABLE 19. Estimated Area of Green Crop or Roots Required for Folding 100 Sheep (or 100 Ewes with their Lambs) for One Month (30 days), assuming that approximately three-quarters of the Crop is Utilised

Type of crop	Gross yield (ton/acre)	Assumed utilised yield (ton/acre)	Area of crop required per month per 100 sheep or per 100 ewes with their lambs:		
			(a) sheep eating 10 lb each/day	(b) sheep eating 15 lb each/day	(c) sheep eating 20 lb each/day
Stubble catch crop—poor	5·0	4·0	3·3	5·0	6·7
Stubble catch crop—average	7·5	5·5	2·4	3·6	4·9
Stubble catch crop—good	10·0	7·5	1·8	2·7	3·6
Stubble catch crop—very good or light kale crop	12·5	9·5	1·4	2·1	2·6
Swedes or swedes and 1000 head kale	15·0	11·5	1·2	1·7	2·3
Swedes or swedes and 1000 head kale	20·0	15·0	0·9	1·3	1·8(d)
Swedes and 1000 head kale	25·0	19·0	0·7	1·0	1·4(d)

Probable consumption per sheep: (a) Fattening hoggetts or in lamb ewes on *restricted* diet of roots; (b) Fattening hoggetts on heavier ration or ewes with lambs on limited diet of roots or green crop; (c) 'Down' type ewe with lambs where roots or green crop form a large proportion of the diet;

(d) An acreage allowance is frequently inapplicable to crops of 20 tons per acre and over as wastage is likely to be very heavy under wet conditions (also see text).

long and narrow, as opposed to square, six feet of fence frontage per cow should be fully adequate for dehorned cattle; allow nine feet for horned animals. For dairy cows the electric wire should be moved daily to provide a fresh feed.

Siting and moving the electric fence wire sometimes poses a problem, especially in tall crops. Solutions to the problem include cutting a lane for the wire by hand which is laborious, cutting the kale with an old mower and then allowing the stock to drag it through the fence or partially knocking the crop down with a bar mounted at the side of the tractor.

Shorter stemmed varieties such as Maris Kestrel, later sowing and close drilling (seven inch rows) help to give a crop which is not so tall and is thus easier to utilise. Alternatively, periodic rows of kale may be substituted by white turnips—one or both of the end coulters of the drill may sow turnips. This practice gives the further advantage of a mixed crop but crop yields vary and the rows of turnips do not always coincide with the appropriate position of the fence. It is essential that the wire is tight and that it is not shorting on the kale leaves.

Crops grown for folding by dairy cows should be sited as near to the farm buildings as possible, say within 300 yards. Long walks are highly undesirable and may result in cattle expending more energy than they obtain from the crop. Except in the mildest areas it is probably safer to rely on stored or conserved fodder for cows after Christmas, if for no other reason than the risk of icy roads or frozen land and gateways.

Fat Lamb Production

The management of the fold is particularly critical with early fat lamb production, when a good supply of ewes' milk and later, of succulent young green stuff, is essential to achieve the rapid growth of the lambs required by this trade. Concentrates are of secondary importance and can only be justified where very early 'forced' fat lambs are wanted.

It is usually advisable to retain the ewes and newly born lambs on a sheltered grass field for 7–10 days before they are moved out to the fold, which ensures that the couples settle down and there is less chance of mis-mothering; the lambs are also stronger and better able to withstand adverse conditions. The provision of adequate shelter or windbreaks is particularly important in exposed situations; lengths of canvas or sacking may be tied to the wire netting or crosses of straw bales one bale high may be laid out in the run-back area.

Forward creep grazing

Forward creep grazing is particularly beneficial and usually improves the growth rate of the lambs which are allowed to creep forward into the next fold and to select the palatable young shoots before the ewes are turned in; the creep may also give the lambs a drier and more sheltered lie. Good hay should be provided to balance succulent green crops.

Heavy crops of roots are unsuitable for folding ewes and lambs in wet weather; apart from the very high waste, the ewes' udders are likely to become excessively soiled with mud, the lambs will not suck and their growth rate is consequently reduced. Losses of lambs may be very heavy if young lambs have empty stomachs and no shelter.

Heavy crops of swedes are perfectly satisfactory for in-lamb ewes which are to be lambed on to grass in the spring, provided that their intake of swedes is restricted to some 10 to 12 lb of swedes per day and that they have a run-back on to some grass—a field of autumn grown grass which is allowed to stand the winter is ideal for the purpose.

No absolute guide can be given as to the size of pen and frequency with which it is necessary to move sheep to a fresh fold. Small pens and frequent moves give the most efficient utilisation but require more labour. In practice the frequency of the move depends on the class of stock and the current rainfall. Ewes producing early fat lambs are best moved at least every two or three days; some shepherds move their sheep daily.

Conversely, in-lamb ewes on a restricted diet may only be moved once a week, although this practice tends to waste roots in frosty weather; once part of the bulb is bitten away, swedes and turnips tend to lose their frost hardiness. The fold should be moved every two days in frosty spells. More frequent moves are required in wet weather when crops become soiled quickly; stock will not eat mud soiled crops.

The greatest economy in labour and fencing costs combined with efficient utilisation is most likely to be obtained by using large mobs of sheep to eat off a substantial area of crop in a short time. This course is more suited to the larger flocks and farms and presupposes an adequate standard of shepherding and large enough handling facilities for the number of sheep. Small flocks and small fold areas tend to be extremely wasteful of labour and other resources.

STORAGE OF ROOTS

Stored root crops are preferable in situations where the land lies too wet in the winter for the crop to be utilised in situ or in areas where prolonged spells of hard frost are of frequent occurrence; a reserve of stored roots for outwintered stock is a valuable insurance against cold snaps in almost any area. The heavy manual labour

F

traditionally associated with root growing, storage and feeding is no longer essential; hand work can be removed almost entirely by the adoption of precision drilling, herbicides, mechanical root harvesters and self-feed hoppers.

Roots may be stored either in buildings or in outdoor clamps; buildings are generally more convenient, easier to work and require less labour. Practically any open-sided building such as a dutch barn, implement shed or galvanised iron lean-to may be used provided that it is waterproof, is dry underfoot, is readily accessible and there is adequate headroom (normal farm trailers require 11–12 feet, large double wheeled trailers require 15 feet minimum headroom when tipped). The roots must also be adequately protected from frost.

Excellent insulation is provided by walls of straw bales supported by sheep netting or galvanised iron and reinforced by old railway sleepers or cheap farm timber. In cold situations, especially if exposed to the north and east, a double wall of straw bales is desirable. Loose straw should be packed in any cavities left between the bales.

The depth of the roots in the store should not exceed 6–8 feet. The roots should be covered with a generous layer of loose straw ($1\frac{1}{2}$–2 feet thick), which is held in place by cheap string netting. The straw must be thick enough to protect the roots against frost and also to prevent condensation of water vapour given off by the roots occurring in the clamp; if condensation occurs in the top of the roots the straw is not thick enough and more should be added.

Types of outdoor clamp vary greatly; the depth does not usually exceed 4–6 feet. In milder or sheltered situations a liberal covering of straw topped over by some two feet or so of hedge trimmings makes a cheap and effective clamp.

In colder areas, earthing up is generally necessary; the colder the situation, the thicker should be the earth covering. Whatever the type of clamp, it is essential that the straw adjacent to the roots is dry and that it is protected from rain. Wet straw freezes readily and is a poor insulating material.

The covering of straw must be adequate and is usually about $1\frac{1}{2}$ feet thick. The correct thickness may be gauged by pushing the tines of a fork into the straw. If the tips of the prongs touch the roots the straw is not thick enough and more should be added. The straw may be kept dry and the thickness of the earth cover reduced somewhat by placing plastic sheeting but the ridge of the clamp must be left open for moisture to escape.

The plastic sheet is best held down by an earth covering; in sheltered situations or with a southerly aspect an inch or two or even periodic spadefuls of earth may be perfectly satisfactory but in cold and exposed conditions a covering of up to six inches is usually to be preferred, especially on the parts of the clamp exposed to the north and east.

The clamp should never be left open in frosty weather; the roots should be re-covered by straw immediately after use. Unwanted straw should be cleared up as consumption of the clamp proceeds; this course prevents roots being left behind under the straw.

Roots are generally topped for storage but the crowns should remain intact. Crops of mangolds with small tops are sometimes clamped with their tops intact; the leaves shrivel and disappear during storage. Crops with heavy tops must be topped before they are stored or clamped. Only roots which are dry, sound, disease-free and which have not been exposed to hard frost should be clamped or stored or they will not keep satisfactorily. The roots should also be reasonably free from dirt.

Self feeding of roots

Roots may now be 'self fed' to stock in a way comparable to the 'self feeding' or 'easy feeding' of silage. One such method is embodied in the ADAS-Walcot Feeder (Fig. 1) which allows swedes to be fed to fattening cattle with minimal labour requirement and little capital expenditure.

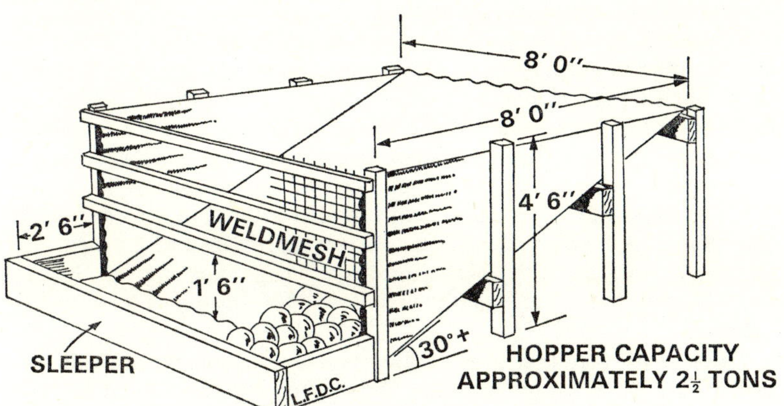

FIG. 1. The ADAS/Walcot Self-feeder.

Published by kind permission of Mr. J. M. Evans, Walcot, Lydbury North, Shropshire, and kindly supplied by Mr. J. Rhodes, Ministry of Agriculture, Fisheries and Food, Shrewsbury, Shropshire.

When access is allowed for 24 hours per day a 6–8 cwt bullock needs about 4 inches of feed face and will consume about 35–45 lb of whole swedes per day. A yard of 20 such cattle would require a hopper 7–8 feet wide. If the hopper holds 2½ tons of swedes it would need to be refilled roughly once a week.

The Walcot Feeder can be constructed without difficulty on the farm and costs approximately £12.00 for materials. The following materials and labour are required:

6 sleepers @ 50p each	£3.00
4 × 9 ft galvanised sheets @ 15p per foot	£5.40
10 × 8 in coach bolts @ 14p	£1.40
3 in and 4 in nails	
Old fencing rails	
Split larch poles	£2.20
Approximately	£12.00

Labour 2 men @ 10 hours = 20 hours.

(Details supplied by Mr. J. Rhodes, DAO, ADAS, Shropshire).

CHOICE OF CROP

The choice of root and brassica green forage crop is limited by the period of the year in which the crop is required for feeding, by the climatic conditions of the district in which it is to be grown and by the length of the available growing season, as influenced by sowing date; the latter is in turn dependent on the harvesting date of the previous crop. The cropping situations in which the various root and fodder crops may be taken to provide feed at different times of the year are summarised in Appendix I.

When the previous crop is harvested early and the fodder crop can be sown in good time, as after early potatoes, the choice is wide; crops such as the *B. oleracea kales* (marrow stem, Maris Kestrel or thousand head kale) which give a heavy yield but require a relatively long growing season are an appropriate choice. The longer sowing is delayed, the more restricted the choice of crop becomes.

Thus, with late sown catch crops drilled after a cereal crop, only the more rapid growers, such as the forage rapes and turnips, are generally suitable. Possible alternative crops are rye or Italian ryegrass but these are usually unsuitable for production of *winter* forage.

There is no problem in providing fodder for consumption in the autumn or early winter months. Quick growing crops such as rape and stubble turnips are perfectly satisfactory provided they are used before the onset of hard winter weather. Thereafter stored crops or crops with an adequate degree of frost hardiness are essential.

Frost hardy crops such as swedes and thousand head kale are perfectly satisfactory provided they can be sown early enough. The difficulty with stubble catch cropping is to find a crop which combines rapid growth with a good level of frost resistance.

Maris Kestrel, thousand head kale and the hardier varieties of yellow turnips will produce a modest crop if sown before the end of July after winter barley in the more favoured situations in the south of England but are very unreliable when sown so late elsewhere.

Forage rape is used to produce spring keep in south Devon but even there it is risky.

The only crops which combine a reasonable degree of winter hardiness with a useful yield from late sowings are hardy green round turnips and the rape kales (hungry gap kale and rape kale); these crops should replace other varieties of turnip and thousand head kale for sowings made from late July to mid-August. The rape kales also flower later than thousand head kale and are occasionally useful for the production of late sheep keep in situations where the grass starts late. However, Italian ryegrass is often preferred for this purpose.

Frost hardiness is influenced to a marked degree by manuring and management. Heavy dressings of nitrogen tend to have an adverse effect and should be avoided with crops grown for consumption in the late winter. Late sown crops tend to exhibit somewhat greater frost hardiness than the more mature early sown crops. The later sowing of the less hardy types, for example July-sown marrow stem kale and September-sown turnips, can certainly improve hardiness but this technique must be regarded as 'second best' to sowing more hardy types and varieties.

Chapter 5

GRAZING CEREALS AND ITALIAN RYEGRASS

ITALIAN RYEGRASS, rye and winter cereals grown for grain may be used to extend the grazing season on grassland and arable stock farms alike. Italian ryegrass is also valuable for intensive grass production during the normal grazing season. Intensive management of these crops has been shown to result in high stocking rates with little or no reduction in the area available for cash crops.

For out of season grazing it is necessary to select a suitable site and to give the crop special management. Fields for early and late bite must be really well drained and not liable to poach unduly. In mild conditions it is technically feasible to produce the keep but the field may be so wet that utilisation is virtually impossible. Well drained light soils warm up and start growth earlier. A sheltered site, preferably with a southerly aspect, is also desirable.

Soil temperatures in the early spring and autumn are not high enough for the nitrifying bacteria in the soil to release the mineral nitrogen necessary for plant growth. For early and late bite, therefore, the whole of the plants' nitrogen requirements must be satisfied by the application of fertiliser nitrogen, which should be applied generously. Plant growth is also limited by low temperatures so that it is quite pointless putting on fertiliser when the temperature is too low for growth to occur. If nitrogen is applied too early or too late in the year there is little or no response and the bulk of the application finds its way into the nearest drain or ditch.

THE CEREALS

Grazing may be produced from cereals in three distinct ways—

I The cereal, which is almost invariably forage rye, is grown specifically for fodder only; it does not provide a grain crop afterwards.

II Cereals, usually wheat, oats or barley are grown deliberately as a dual-purpose crop, to provide both an early bite and a crop of grain afterwards, and

III A 'casual' grazing may be taken from cereals grown primarily for grain. This occurs either because the crop has become winter proud or because there is a shortage of fodder.

Cereals Grown Specially for Grazing

Rye

Owing to its capacity to make good growth at low temperatures, rye is particularly suited to the provision of really early spring grazing in the 'hungry gap' which occurs before any other grass or cereal starts to grow. This is its main advantage. Rye is not subject to the soil-borne diseases of wheat, oats and barley, and does not appear to perpetuate them.

Rye suffers from a number of disadvantages. The stem develops very quickly; if the crop is not eaten in an early, leafy stage, the stems become course, fibrous and unpalatable and are then neglected by stock, with consequent heavy wastage. The period during which the crop is in the optimum stage for grazing is therefore short.

Apart from its earliness in starting growth, rye compares very unfavourably with Italian ryegrass, which is far more versatile, has a much longer useful life, is more productive and has a lower seed cost. The cost of cultivation for a given amount of feed from rye is also greater because of its shorter life. Italian ryegrass is also much easier to conserve in the case of surplus.

The main use of forage rye is for the provision of extremely early grazing and it is well suited to the really intensively managed dairy farm to provide the first green bite of the spring. Forage rye is also invaluable in an emergency, such as a severe shortage of conserved winter fodder and in cold areas where growth starts late.

When grazing rye with sheep, it is better to take two relatively short grazings rather than try to obtain a single heavy grazing. The liveweight gain of lambs fed on rye can be as high as 650 lb per acre with intensive stocking.

Varieties

The best policy is to use varieties of rye bred specially for fodder production. These comfortably outyield the grain varieties and the old common rye which was used before the introduction of special forage varieties.

The following have been found to be the most suitable for British conditions: Bernberg, Lovaszpatonai, Greenfold and Rheidol. There is little difference between them in earliness, yield of green fodder and recovery after cutting or grazing. Early grazing may also be obtained from the grain varieties of rye when the crop is grown occasionally for grain; a side effect is to reduce the length of the straw.

Sowing date

If it is to produce a really good flush of early growth in the spring, rye must be well established and tillered; early sowing is essential. The best time is in August or early September, the earlier date being the most suitable for colder areas and where more autumn grazing is required. Even in the relatively mild south west of England sowing should, if possible, be completed by mid-September. October sowings will produce a satisfactory yield of grain but the yield of early bite, which requires quite different management, is usually disappointing. Late sowing does not give the necessary flush of early growth and the yield can easily be reduced to a mere third or less of that from sowings made at the proper time.

Rye must not be allowed to enter the winter in a 'proud' condition. Autumn growth should be grazed off during November to ensure the development of numerous short sturdy tillers. Once grazing has been completed the stock should be removed and the field rested until the early bite is ready.

Seed rate

A fairly heavy seeding, approximately $1\frac{3}{4}$ cwt (200 lb) per acre, is necessary. The inclusion of Italian ryegrass is of doubtful value unless it is intended to provide a series of further grazings after the early bite. If so, the seed rate of the rye should be reduced to 1 cwt (112 lb) and 20 lb per acre of Italian ryegrass is then included.

Manuring

Although rye grows well on poor light soils and is virtually insensitive to soil acidity, generous manuring, particularly with nitrogen, is essential for satisfactory production. About 30 units each of P_2O_5 and K_2O are applied to the seedbed according to soil analysis. If November grazing is intended, 30 units of nitrogen should be included. Nitrogenous top dressing is essential for the early bite; 40–50 units of nitrogen per acre should be applied from the end of January to late February according to the district.

The spring application should be preferably staggered in order to provide a succession of feeds.

When further grazings are required, which is always the case when Italian ryegrass is included, further nitrogen should be applied immediately after each grazing.

Grazing Autumn-Sown Cereals

'Dual purpose' crops

The feasibility of growing winter cereals for the dual purpose of providing an early grazing and then a grain crop is limited to those parts of the country with mild winters and early springs, notably west Wales and the south west of England. In these areas early spring growth, with consequent 'winter proudness', is commonplace and provided sowing does not occur too late in the autumn, early feed from autumn-sown cereal crops can usually be relied upon. Apart from the additional fodder supplied, spring grazing of forward cereals is often highly beneficial in the south west of England, as it reduces straw length, the tendency of the crop to lodge and probably some of the risk of plant disease too. The practice of removing winter proud growth also permits earlier sowing, allowing a farmer to plant his crop when soil conditions are satisfactory; he is then sure of sowing his crop, which is not always the case when sowing is delayed until late in the autumn or even early in the new year.

Emergency grazing

The further north and east one travels, the less likely are autumn-sown cereals to become winter proud or to provide spring grazing. Cold winters prevent this. Winter cereals are then usually grazed only as an emergency measure, either to remove excess growth if the crop has become too forward or when there is a shortage of other keep.

Management of winter cereals for grazing

Whether spring grazing of winter cereals is specifically planned or not, the underlying principles are the same. The amount of keep available in the spring depends on the date of sowing and the temperature during the winter and early spring. Early sowing, either in late September or early October, results in a more forward crop, which is likely to give more spring grazing.

A little additional nitrogen in the seedbed, say 12 to 15 units nitrogen, also brings the crop forward. However, if it is not intended to graze, such a combination of nitrogen and early sowing is usually a mistake. Acreages of winter cereal treated in this way should be limited.

The amount of grazing on offer at any given date varies greatly from one year to another. Thus, after a mild winter, winter oats grown at the Welsh Plant Breeding Station become winter proud and gave 1,200 lb dry matter per acre in mid-March. After a more severe winter in the following year, only 620 lb dry matter per acre was on offer at the same date. In the first instance, grazing was necessary to prevent lodging and when properly carried out actually improved the yield of grain. However, only in very winter proud crops does grazing normally improve yield. The yield of straw is reduced progressively as grazing is delayed.

The stage of crop growth at which grazing occurs is critical. Grazing must be completed before the growing points of the cereal plants start to extend and there is any likelihood of removing the embryo ears. Should this occur, the loss of grain can easily reach 10 cwt per acre; in really bad cases losses of up to 87 per cent of the grain yield have been recorded experimentally.

In practice, the arrival of the critical stage after which the crop should not be grazed may readily be determined by examining a few plants before grazing. If the growing point at the centre of the ensheathing leaves has not started to elongate, then the crop is still fit to graze. Once elongation has started, the embryo ear is likely to be removed during defoliation and further grazing is almost certain to reduce yield. The date on which this stage is reached can differ markedly from one season to another.

When the early spring is warm, the flower buds are laid down early; crops which are already winter proud should therefore be grazed early and relatively leniently. Less forward crops may be grazed later and more severely. In general, the safest policy is to err on the side of grazing a little early. Grazing winter cereals should normally be finished by the end of March or the first week in April. The stage of growth at which grazing occurs is more important than the date.

Top-dressing with nitrogen before grazing increases the amount of grazing but is inclined to reduce the yield of grain. Once they have been grazed, cereals respond well to a nitrogenous top dressing. This should never be omitted with grazed crops; from 40 to 50 units of nitrogen can usually be justified, although the precise amount depends on soil fertility, species and variety of cereal.

Wheat and shorter, stiffer, strawed varieties of oats may need the larger amount. The grower should be governed by local experience. However, it should be noted that nitrogenous top dressing does little to offset the harmful effects of excessively late grazing. Winter proud crops which remain ungrazed do not give a worthwhile response to an application of nitrogen.

ITALIAN RYEGRASS

The various forms of Italian and related short-lived ryegrasses are identical in their outward appearance but differ in their behaviour and agricultural characteristics. Broadly they include three distinct types:

I Westerwolds or westerwolths ryegrasses.
II The 'true' Italian ryegrasses.
III The hybrids between Italian and perennial ryegrasses.

The earliest flowering types of perennial ryegrass such as Aberystwyth S24, are sometimes included with the short-lived ryegrasses, but perennial ryegrass is botanically quite distinct from Italian ryegrass and usually irrelevant to leys of less than one year's duration and to catch cropping. All the short-lived ryegrasses establish extremely rapidly and recover quickly after cutting or grazing. If properly managed they will provide a very large amount of fodder during their relatively short life.

Westerwolds ryegrass

Westerwolds ryegrass is strictly an annual, capable of flowering within about three months of sowing. It is the only grass, other than the cereals, which will produce a cut of hay in the year in which it is sown. Its use is limited entirely to situations where high productivity, largely from cutting, is required within three to six months of sowing and should always be sown in the spring or very early summer.

Westerwolds is usually not winter hardy and is therefore quite unsuited to autumn sowing or for use as the main constituent in leys of over six months' duration. Although once popular for sowing after early potatoes, Westerwolds ryegrass is now of little importance in the United Kingdom. Italian ryegrass, which is winter hardy, is preferable in almost all situations.

If Westerwolds is to be used, leafy bred varieties, which are mostly of Dutch origin, are far superior to the stemmy short-lived 'commercial' Dutch type, which is also of much lower productivity. The following varieties are the most suitable: Barenza, Million, Motto and Sceempter. There are also tetraploid varieties.

The 'true' Italian ryegrasses

'True' Italian ryegrasses are biennials or short-lived perennials and do not flower in the year of sowing. Persistency differs with variety, the most persistent varieties living for up to 18–24 months, depending on the management they receive. As a sole constituent, this type is usually restricted to use in leys of not more than 12–18 months' duration. These 'true' Italians constitute by far the most important group.

The hybrids between Italian and perennial ryegrass

Hybrids combine some of the out-of-season growth of Italian ryegrass with some of the persistency and sward density of that of perennial ryegrass. Some varieties are virtually Italian ryegrasses while others approximate more closely to the perennial parent. The varieties in this group are mainly used in leys of at least 18 months' duration.

Uses of Italian ryegrass

Italian ryegrass has a wide application on either grass or predominantly arable farms and is much more versatile than rye. It is relatively cheap and easy to establish and produces keep some 4–6 weeks after sowing on bare ground; these are most useful attributes for catch cropping and for use in emergency.

The outstanding feature of Italian ryegrass is its capacity to produce abundant 'out of season' early and late bite. Spring growth starts before that of any variety of perennial ryegrass and continues later into the autumn, extending the grazing season considerably. At the same time heading occurs some two weeks later than in the earliest varieties of perennial ryegrass, so that with the quick recovery from grazing there is no difficulty in producing an early bite followed by a heavy and leafy cut of silage or hay.

Italian ryegrass is also particularly suited to really intensive grazing management with either sheep or cattle.

The possibilities of Italian ryegrass therefore fall broadly into two categories:

(i) For use as a short term catch crop.

(ii) As a ley from one to a maximum of two years' duration. There are various intermediate stages.

When grown as a catch crop, Italian ryeglass may be undersown in cereals or sown on bare ground after an early harvested crop such as early potatoes to provide either late summer and autumn keep, cheap over-wintering of ewes (supplemented by other green crops) or early bite. The 8–9 month ley is both popular and highly productive. Sown in a cereal stubble it provides some autumn grazing, excellent early bite and a heavy cut of silage and yet gives ample time for a sizeable crop of kale or rape to follow. If the next crop should be swedes or vegetables, it is better to be satisfied with one or two grazings and prepare the ground in good time for the succeeding crop.

With heavy nitrogenous manuring and intensive stocking, of the order of eight ewes and their lambs to the acre, the output of Italian ryegrass leys of 12 months or longer duration can be extremely high. These leys are equally suited to intensive dairy farms and arable farms with a sheep flock. In the latter case, they can be used to replace the traditional red clover ley and act as a profitable fertility-restoring 'break crop'.

Varieties

Varieties differ considerably in their characteristics and in their relative yields at different times of the year. Of the varieties recommended by NIAB, choice is governed by the location and duration of the crop and the availability and cost of the seed (Appendix 2).

In areas with cold winters such as the north and east of England and in Scotland, winter hardiness is of prime importance; the hardier varieties of Continental origin should always be chosen.

For short-term catch crops persistency is unimportant and the shorter lived varieties are quite suitable and are usually somewhat cheaper. For leys of longer than 12 months' duration leafy persistent varieties are necessary. A few pounds of S24 Perennial ryegrass may also be added as an insurance against loss of plant.

Date and method of sowing

Italian ryegrass may be sown at any time from early spring to early autumn, the precise limits depending on the climate. In cold districts sowing should preferably be completed by mid-August, while in the milder parts of the south and south-west of England, sowing in early September is highly satisfactory; in fact sowing can occur here in the latter part of September. While later sowings may succeed, they are inadvisable as they carry a much greater risk of failure to establish satisfactorily before the onset of winter; the yield of early bite available during the following spring is also reduced.

Sowing date is substantially determined by the period in which the grazing is most required. For early bite in the spring, sowing late in the previous summer or early autumn gives the earliest grazing and best yield; a clean cereal stubble provides the ideal opportunity. Second year crops or crops sown the previous spring are later in starting growth and much less satisfactory for the really early bite.

On the other hand, spring sowing is markedly superior where midsummer grazing is the prime consideration. Sown in March or early April without a cover crop, Italian ryegrass will produce a wealth of leafy grazing free from flowering heads right through the summer of sowing. No flowering heads are produced because the plants have not passed through a winter cold spell to 'vernalise' them, which is essential for flowering to occur in this type of grass. Valuable late summer and autumn grazing may be obtained by sowing after early harvested crops such as early potatoes, peas, cereal silage and in some areas winter barley.

When maximum autumn keep is the prime consideration, sowing

must occur early enough to allow adequate plant development to occur. This requires either sowing by midsummer without a cover crop or undersowing in a cereal crop the previous spring; a practice favoured by a number of arable sheep farmers. Sowing in a spring cereal stubble after harvest is too late for full autumn production.

In a single comparison in 1970, the author obtained 484 sheep grazing days per acre between harvest and November 30th compared to only 103 days from a crop sown in a spring barley stubble on September 4th. Spring and winter sown cereals or better still, whole crop cereal silage are suitable for undersowing. Legumes or cereal/legume mixtures are unsuitable as they tend to produce a dense cover which may smother the ryegrass seedlings.

There are certain risks attached to undersowing Italian ryegrass. In wetter districts excessive growth of the ryegrass in the base of the corn may give rise to difficulties in harvesting. Vigorous ryegrass growth may also reduce the yield of grain, while on the other hand there is the possibility of the ryegrass being smothered in a dense crop.

Moderate cereal seedrates and nitrogen application should be the rule when undersowing. Although undersown Italian ryegrass gives superior autumn keep, its spring growth is less vigorous than sowings made without a cover crop in the previous autumn. Undersowing should generally be confined to situations where Italian ryegrass is being grown specifically as a short term catch crop to give an immediate changeover from corn and provide autumn and early winter grazing or for overwintering stock cheaply on arable land.

Seed rate

The optimum seed rate depends on the circumstances. When sowing without a cover crop, it is essential that a thick sward should be developed with the minimum delay in order to suppress weed growth and reach full productivity as quickly as possible. The seed should then be sown relatively thickly; from 20 to 25 lb per acre is usually necessary with broadcasting; less than 20 lb per acre is false economy. With adverse conditions, such as very early or late sowings or a somewhat indifferent seedbed, up to 30 lb may be preferable, although somewhat expensive. If the seed is drilled 18–20 lb per acre is usually adequate. When undersowing in corn, there is not the same urgency to form a sward so quickly; seed rates as low as 12 lb per acre are usually adequate, provided the seed is sown evenly and immediately after the corn on a really good seedbed.

For broadcasting under winter corn, with less accurate seed distribution, or where the tilth is not so satisfactory, 15–18 lb per acre is probably safer for undersowing. The purchase of 'clipped' seed, when the awns are partially removed, helps to eliminate 'bridging' in the seed hopper and ensure a better flow of seed in the drill.

Italian ryegrass may also be grown in mixture with rye, rape or turnips. Any of the following may be added to 20 lb Italian ryegrass:

1 cwt forage rye

or ½ lb hardy yellow turnips

or 2 lb forage rape.

Seedbed preparation and sowing

Italian ryegrass may be direct drilled or sown on a prepared seedbed; in the latter case the seedbed should be fine, firm and level. Provided that the soil is free working, well drained and there is no pan, depth is unimportant. Ploughing is only desirable when there is enough trash present to block the cultivating implements or in preparation for early spring sowing when autumn ploughing generally produces a good frost mould, helps surface drainage and allows early sowing on a good firm seedbed.

In contrast, when Italian ryegrass follows early potatoes, the land normally only needs a pass or two of the harrows to level it before sowing.

A complete burn-off is the ideal preparation for sowing out cereal stubbles but this involves sacrificing the straw. When the straw is required, as short a stubble as possible should be left and then sprayed with paraquat. The seedbed may then be prepared with 'minimal cultivations' (Chapter 7).

When undersowing spring cereals or spring sown cereal silage, the best takes are usually obtained by sowing the grass immediately after the corn; if possible on the same day. If the risk of the ryegrass making excessive growth in the corn is considered to be too great, the sowing of the grass may be delayed until after the corn has emerged and is reasonably well-established.

With winter cereals there is a risk of poor establishment because of severe shading from an advanced cereal crop or lack of an adequate seedbed. Winter cereals are best undersown before full spring growth starts, while forward crops can usefully be grazed. Enough tilth should be produced to cover the seed, which may be either cut into the soil with a disc drill running across the cereal rows or broadcast and covered with one or two strokes of the heavy harrows and then rolled in.

Traditionally, winter wheat stands up to these cultivations best but operations must be completed early enough to prevent damage to the cereal and give the grass the maximum time to establish before competition for light and moisture from the cereal becomes really severe.

When there is adequate rainfall and soil moisture for establishment, broadcasting the seed gives highly satisfactory results and is the quickest method. Broadcasting on a ring-rolled surface, followed by harrowing and a further rolling, gives good coverage of the seed, although on soils where the surface is inclined to cap it is preferable to finish off with a chain harrow and not leave the surface too tight. However, the seedbed should still be firm.

On the other hand, drilling is a good deal safer under dry conditions. The seed is cut into the soil moisture and positively covered, resulting in much more rapid establishment than with broadcasting. The shallower the seed can be drilled the better.

There is everything to be said for drilling in preference to broadcasting with spring sowings planted without a cover crop and when undersowing winter cereals. All too often drought, especially in the spring, results in a slow or patchy establishment from broadcast sowings. Drilling also results in some economy in seed usage: only about two-thirds of the broadcast rate need be sown. The efficiency of the drill is unaffected by windy conditions.

Few farms possess either special grass drills or corn drills with close spacing and it is usually necessary to use a corn drill with coulters spaced seven inches apart. Drilling two ways, if possible diagonally rather than at right angles, gives the most even distribution of the seed. Even with clipped seed, Italian ryegrass may be inclined to bridge in the drill and consequently require dilution. A few pounds per acre of an inert diluent, mixed with the seed should help its passage through the seed box.

With a combine drill the seed may be bulked with up to 100 lb per acre granular superphosphate and sown through the manure hopper. On no account should either the amount of superphosphate suggested be greatly exceeded, or concentrated compound fertiliser be substituted. Even at relatively low rates of compound there is a considerable risk of impairing the germination of the seed.

Manuring

Italian ryegrass is essentially a grass for high fertility, requiring an adequate supply of soil moisture and the liberal use of nitrogenous fertiliser. Under really intensive management a good Italian ryegrass sward can use 300 units per acre with the utmost profit in a full year. Without the necessary liberal manuring and high fertility the performance of this grass is positively wretched, while the omission of nitrogen in spring and autumn means that growth is virtually non-existent at these times.

It is best to apply phosphate and potash fairly liberally for crop establishment and work the fertiliser into the seedbed. Some 50-60 units of each per acre is usually worthwhile but the precise amount depends on soil type and analysis. It is at this stage that Italian ryegrass gives its greatest response to phosphate. Crops sown without a cover crop should normally receive 40–60 units N in the seedbed, while undersown crops should not receive nitrogen until the corn has been cleared.

After the initial application of fertiliser for establishment, manuring depends on the management employed. When the ley is solely grazed and the stock remain in the field all day, which is the usual case with sheep, all fertiliser residues are returned to the soil via the excreta and only nitrogen need be applied after each grazing.

In contrast, when a cut for hay, silage or zero grazing is taken, substantial quantities of minerals, mainly potash, are removed and a fertiliser containing potash should be given before and after each cut. If all the potash is put on before the cut, luxury uptake occurs; the whole dressing of potash is then taken up but the additional potash gives no extra yield. The regrowth or succeeding crop may even suffer from lack of potash.

The greater the weight of grass removed, as influenced by the size and frequency of the cuts, the greater will be the need for subsequent applications of potash. The maximum response to phosphate is about 50 units per annum but with cutting only, potash must be applied in proportion to the nitrogen; approximately 0.8 units K_2O per unit of N. Even if the yield of the ryegrass is not reduced, the application of insufficient potash may still result in a shortage of potash for succeeding crops.

The amount of nitrogen that should be applied before each cut or grazing depends on the interval likely to occur between the application of the fertiliser and the removal of the grass. As a general guide, two additional units N per acre are necessary for every day the length of the interval increases. Thus, if there is an interval of 20–22 days between grazings 40–44 units of nitrogen should be applied.

If more than 2 units N per day is applied in a single dressing the fertiliser is taken up by the plant but is not translated into an increased yield of grass. A cut of silage requires a longer interval for growth; 60–70 units N per acre may then be worthwhile. Precise details depend on the purpose for which the ryegrass ley is used.

Management

The initial management of the ley should be designed to promote maximum tillering, with the consequent thickening of the sward as quickly as possible. As soon as the corn is removed from undersown

G

leys or there is enough grass to provide keep for a day or two where there is no cover crop the stock, preferably sheep, should be brought in. The cloven hoof consolidates the ground and this, combined with the removal of the leading shoot, helps to induce rapid production of tillers. The stock should be followed immediately with a dressing of about 40 units N and the sward rested until the next grazing.

Except when outwintering stock on an Italian ryegrass ley that is to be ploughed, grazing management should always be intensive. If the grass is allowed to become too long, especially with sheep, the unpalatable bases of the shoots are neglected and waste is heavy. Furthermore, the vigour of the sward is irretrievably lost. With sheep in particular it is better to err on the side of grazing too frequently.

It is better to have two short grazings than a single longer bite. On and off grazing produces the best results, with nitrogenous fertiliser applied after each grazing. The key to success with this grass is really dense stocking, heavy manuring and tight grazing control. The acreage sown should therefore be restricted to an amount which can be managed efficiently.

As first year stands are the earliest and most productive, there should be a regular sequence of planned sowings. Unless an Italian ryegrass ley is in really good condition it should not be left down longer than 12 months. Many farmers prefer to grow this grass as an 8–9 month catch crop.

To produce late bite, some 40 units of nitrogen should be applied for each grazing; in the milder parts of south west England, nitrogen may be profitably applied until the end of the first week in October with an open autumn but in colder areas, the response after late August or early September is likely to be disappointing.

Undersown leys should be grazed and receive their first application of nitrogen as early as possible, as this determines the amount of grass they will provide.

All Italian ryegrass leys which are required to provide early bite in the spring should be grazed off closely by late November and then rested until the stock are turned in during the early spring. If Italian ryegrass is allowed to enter the winter in a proud condition and is then subjected to hard frost, winter hardiness is almost non-existent and the sward is usually destroyed completely.

Conversely, the small amount of regrowth that may occur before Christmas must not be stocked. Straight nitrogen at the rate of 60 to 70 units per acre is usually applied in February or early March according to season and district.

Owing to some risk of grass staggers from early spring applications of potash, it is safest to delay the compound fertiliser until the second application. For a cut of silage 60–70 units N, 30–35 units P_2O_5 and 30–35 units K_2O should be applied as a compound with a 2:1:1 ratio. For grazing 40 units N and a correspondingly smaller quantity of phosphate and potash is needed.

WHEN TURNING OUT TO GRASS

Very early young spring grass is rich in protein and frequently makes cows' dung very loose. Problems with bloat may also occur. Precautions should include gradual introduction of the stock to the grass, avoiding turning out on to wet grass in the early morning, giving a dry feed before grazing and ensuring that the stock have access to sweet hay or straw to provide fibre. Balanced concentrates should be replaced by cereals or low protein dairy cake; the earliest grass is frequently deficient in carbohydrate and may only provide for the first 2-3 gallons.

Hypomagnesaemia, also known as grass staggers or grass tetany, can be a real problem with intensively managed early spring grass and cereals. Acute cases are frequently fatal and prevention is by far the safest course. Death can be rapid and immediate veterinary treatment of affected animals is essential.

Supplementation with magnesia (calcined magnesite) at the rate of 2-4 oz per cow per day is essential; ewes should receive $\frac{1}{3}$ oz per day.

Concentrates are the most satisfactory method of giving the magnesia while cake feeding is required. Purchased or home produced nuts are the most palatable. The magnesia powder can be mixed with rolled cereals; palatability is greatly improved by the addition of molassine meal. The weight of magnesia required for mixing depends on the rate of concentrate feeding; the average daily amount of concentrate consumed per cow should contain two ounces of calcined magnesite:

Concentrate per cow per day	Calcined magnesite per ton
lb	lb
$3\frac{1}{2}$	80
7	40
14	20

Once cake feeding has stopped calcined magnesite can be applied at 40 lb per acre to the grass as a specially prepared fine powder. Very even application is unnecessary and a fertiliser distributor is satisfactory. Several days grazing may be treated in one application.

The animals must receive their magnesia *daily* as they do not build up adequate body reserves of this element. It is also of prime importance that the intake is adequate.

Magnesian limestone, used as a liming material, helps on some soils.

Area required for early bite

Allow $\frac{1}{2}$ acre per cow or cow equivalent.

Chapter 6

CEREALS GROWN PRIMARILY FOR SILAGE

THE IDEA of growing arable crops for silage is not new. Immature crops of cereals and legumes were among some of the first crops to be ensiled when silage making was introduced into England over 100 years ago. Interest lapsed until after the first world war, when a number of enthusiasts erected tall concrete tower silos on the American pattern and proceeded to fill them with either immature maize or mixtures of cereals with legumes, the latter producing the so-called 'arable silage' of the day.

Many of the resultant silages, especially those produced from maize, were excessively wet, poorly fermented, of very low feeding value and most unpleasant to handle. Apart from a few isolated bursts of interest, notably with maize, there has been little interest in growing arable crops for silage until the last few years. Reasons for this lack of progress are not hard to find. Apart from the utter inadequacy of the machinery of the day, the crops which were ensiled were often quite unsuitable, the silos were not airtight and the techniques of silage making were inefficient and frequently little understood.

Arable crops grown for silage must meet certain requirements. They should not be unduly difficult or expensive to grow and their period of harvesting should not clash with that of grass or other crops grown on the farm. If it is to be suitable as a 'break crop' in an intensive cereal growing system, a silage crop should not carry or be susceptible to pests and diseases attacking the cereals grown for grain.

The crop must produce a high yield and at the same time it must have a high enough dry matter content at harvest to eliminate the need for wilting and the likelihood of any loss of dry matter through effluent from the silo. Such a material normally produces a satisfactory type of fermentation.

An adequate dry matter content is also essential to produce a silage which is sufficiently concentrated to permit a high intake of dry matter by the stock consuming it; this requirement is of the utmost importance with the more productive animals such as high-yielding dairy cows and fattening bullocks; at the same time the silage must be palatable and of high digestibility.

When tower silos with automated unloading and feeding systems are used, the silage must be uniform and easily handled by the machinery. The arable crops suitable for silage are maize and other cereals which are used to produce the so-called 'whole crop cereal silage'. Wide commercial experience has been gained with each of these crops but there is yet much to learn. They must therefore be regarded as still in the developmental stage.

MAIZE

Maize is not a difficult crop to grow provided that climatic conditions are favourable, birds are kept away in the early stages and that varieties are chosen which are suited to the purpose for which the crop is being grown. However, it must be emphasised that a very high standard of husbandry is essential for success.

Maize may be grown in England for four purposes:

 (i) Grain.
 (ii) Silage.
 (iii) Green forage.
 (iv) Sweet corn.

The cultural techniques employed in each of these forms of production are identical in many respects, but the end products are essentially quite different and they must be treated as basically distinct crops. The prime interest in this chapter is of course in the production of maize for silage and green forage but limited information is also given on growing maize for grain and sweet corn for growers who may wish to experiment. The information also helps to clarify the differences between the four types of production.

(i) Grain

Only since the introduction of really early ripening varieties capable of producing mature grain in a restricted growing season has the production of grain been commercially feasible in England; even then, the area in which maize can be grown successfully for grain is still very limited and only the earliest ripening varieties are suitable.

Maize, grown for its grain, is becoming an increasingly popular break crop on intensive cereal farms, as it is not susceptible to the common cereal diseases, it is not harvested until the normal cereal harvest is completed and the only additional equipment required is a precision drill and a maize attachment for the combine harvester. Yields of dry grain considerably in excess of two tons per acre can be obtained from suitable varieties. The stover or straw is usually valueless.

(ii) Silage

The object with silage maize is to produce a heavy yield of dry matter per acre with a high dry matter content; ideally 50 per cent or more of the dry matter should be in the form of ear. Yields of $4\frac{1}{2}$ to $5\frac{1}{2}$ tons of dry matter per acre may be obtained, which compare very favourably with those obtainable from intensively managed grassland, especially in hot summers and areas of low rainfall. As the crop is harvested in a less advanced stage of ear development than maize for grain, silage maize may be grown satisfactorily over a much wider area of the country.

Good maize silage is remarkably uniform and is an excellent bulk food for the provision of maintenance and part of the production requirements of fattening cattle and dairy cows. In the latter case any deficiency in protein in the total ration must be made good by feeding additional protein, as maize silage contains only 8–10 per cent protein in the dry matter.

Harvesting occurs in late September or early October and can extend the silage making season if required. The only special items of equipment necessary are a precision drill and a forage harvester with a fine chop and a maize attachment; smaller and cheaper single-purpose harvesters for maize are also available. Maize silage is primarily suitable for the larger grower. Small-scale production is not worthwhile.

(iii) Green Forage

In the case of green forage the prime objective is to grow a large yield of succulent green material, composed largely of leaf and stem, as quickly as possible. In contrast to silage maize a high yield of ear is not required. The seed is sown much more thickly than with the two previous methods and early sowing is much less important; in fact, in Germany quite a substantial acreage of maize is sown very thickly as a catch crop. Extremely heavy yields of green forage have been grown in the U.S.A. and in Europe, although, here, the trend has been for the area of silage maize to increase at the expense of that grown for green forage. Maize is rarely grown for green forage in England.

When sown thickly for green fodder, maize could well be considered to have considerable potential for farmers with large zero grazed herds. However, owing to its deficiency in ear, any maize grown in this way, which is not fed as green forage, makes a very poor raw material for silage. It is therefore usually safer to grow silage varieties and drill at a spacing suitable for silage, so that should part of the crop not be required for green forage, it is perfectly suitable for silage. The area planted specifically for green forage should be restricted to an amount which the grower can be absolutely certain of feeding green.

(*iv*) *Sweet Corn*

Unlike other types of maize the sugar produced by the plant is only converted very slowly into starch. Consequently, when the cobs are picked and marketed in the 'milky' stage, they are very sweet; *only specially bred varieties will produce the sweet corn 'on the cob' essential for this trade.*

Although belonging to the same species, sweet corn is quite distinct from the types of maize grown for grain and silage and on no account should ears taken from these types be sold either as sweet corn or 'corn on the cob'. Although sugar is produced by varieties grown for grain or silage it is rapidly converted into starch; the cobs are therefore starchy and quite unsuitable for the table. Selling cobs from grain or silage varieties can only ruin the reputation of an excellent luxury product.

Growers of grain or silage maize wishing to serve a local market must plant a specific sweet corn variety and drill at an appropriate spacing. There is reason to believe that it is necessary to isolate sweet corn from silage and grain varieties, as the pollen from the latter two can cross-pollinate sweet corn and produce a number of hard starchy grains.

A distance of 200 yards between maize, grown for grain or silage on the one hand and sweet corn, on the other, would appear to be fully adequate. The effect of cross-pollination is likely to be minimised by an increase in the area of the sweet corn, as the outside rows of any crop tend to act as a pollen barrier.

Where maize can be grown

Although maize is widely cultivated in Europe, it is basically a sub-tropical crop and does best in really hot sunny summers. It is very sensitive to low soil temperatures in the spring, makes poor growth in areas with a cool wet summer and is killed by the first sharp frost of autumn. It can therefore only be relied on in the warmer parts of England and in favoured situations which have a reasonably long growing season.

The northern limits of the area in which maize may be grown

depend on the earliness of the variety chosen and on the state of maturity of the ear at which the crop is to be harvested. The more mature the ear required, the further south must the boundary lie. It is essential that the ear has reached the required state of maturity before the first autumn frost brings growth to an abrupt end.

The area suitable for grain production is therefore the most limited and lies to the south and east of an imaginary line drawn roughly from south Norfolk via Somerset to East Devon but excluding land 600 feet or more above sea level. Grain has been successfully harvested in trials at Seale-Hayne and it does not seem unreasonable to suggest that the area suitable for grain production might be extended to take in a strip running parallel to the coast of south Devon and south Cornwall but avoiding high land and very exposed situations.

Silage maize usually grows well in the area south of a line drawn from Bristol to the Wash. Satisfactory crops may be grown as far north as Yorkshire but the further north the crop is grown, the greater the need for really well favoured sites and to plant only the earliest maturing varieties, even if this results in some sacrifice in the total yield of dry matter. It is usually difficult to obtain ears of a satisfactory maturity well north of the accepted silage maize area. The breeding of improved varieties could well extend the potential silage maize area further north.

Sheltered sites with a southerly aspect that warm up early in the season are ideal but by no means essential. Exposed or windy situations, cold slopes facing north, and land over 600 feet above sea level should be avoided, as maize is subject to wind damage and liable to lodging. Lodged crops are very difficult to harvest and likely to result in considerable contamination of the crop by soil.

Suitable soils

Maize can be grown successfully on a wide range of soils but drainage and tilth must be good; deep free working loams are best. Soils that are shallow or are liable to dry out should be avoided, as maize is extremely sensitive to drought when tasseling and silking, that is male and female 'flowering', which occurs in the commonly grown varieties approximately 14 weeks after sowing.

Cold heavy soils are unsuitable as they are slow to warm up in the spring and give very difficult harvesting conditions in a wet autumn.

Place in the rotation

When yield of ear is of prime importance, as with grain, silage or sweet corn production, maize requires a long growing season and must only be grown as a main crop; it is quite unsuitable for growing as a catch crop. In very favourable situations and on free working soils, maize can of course, follow a catch crop such as rye or Italian ryegrass which is grown for early bite, provided that the catch crop is cleared in ample time to allow the production of a really good seedbed. However, this practice is decidedly risky, as the cultivations necessary for seedbed preparation may result in considerable loss of soil moisture and the tilth may also be unsatisfactory.

It is generally much safer to forego a catch crop, particularly in the drier parts of the country and on soils where it is difficult to obtain a good tilth after late ploughing. The absence of a crop in the spring also allows cultivations to take place for the destruction of wild oat seedlings.

Maize usually replaces the root break and is therefore often grown between two corn crops. It is particularly desirable to stubble clean the land as soon as the preceding cereal crop has been removed; this operation is helpful for rotting the stubble and for controlling grassy perennial weeds.

Maize also grows very well after leys or permanent grass, provided the soil is treated against wireworm when necessary. Maize is not subject to the normal cereal diseases. If required it may be grown in two consecutive years, when it provides a very effective double break crop against take-all and eyespot. When the appropriate rate of atrazine is applied to the first of these crops, maize is also a useful cleaning crop for couch, as well as many annual weeds. Winter oats may not follow a spring application of atrazine.

Manuring

Maize does not fail until the pH falls to about 5·2 but it is advisable to see that the land is adequately supplied with lime. The pH of arable soils should in any case not be allowed to drop below 6·0, especially where crops such as barley and sugar beet, which are very sensitive to lack of lime, are grown.

The soil should be in fertile condition with a good organic matter content. Soil fertility is greatly improved by ploughing in a grazed ley rich in clover, well rotted farmyard manure or slurry in the autumn. It is vitally important to apply adequate nitrogenous fertiliser: the usual level of application is 100 units of nitrogen per acre but this depends on whether a dressing of dung has been given and on the fertility of the soil. After a heavy dressing of dung, 80 units of nitrogen may be adequate but the rate of application should be increased to 150 units of nitrogen per acre when the maize is grown as a break crop in a run of cereal crops or after a wet winter or early spring when the available nitrogen in the soil is likely to be low.

The optimal levels of phosphate and potash depend largely on the state of the soil reserves, as indicated by soil analysis. When good soil reserves of phosphate and potash have been built up, 50 units each of P_2O_5 and K_2O are generally adequate. If there is any doubt, heavier dressings should be given, as an inadequate supply of potash tends to encourage lodging.

The entire dressing of fertiliser should be broadcast during the course of seedbed preparation so that it is well incorporated into the soil. If the crop must be sown with a corn drill compound fertiliser should not be combine drilled, as it is highly phytotoxic to germinating maize seed. There is no evidence to indicate any advantage to splitting the dressing of nitrogen. Top dressing may even reduce yield: granules of fertiliser are trapped in the leaf 'funnels' and cause scorch, which may be severe. Should it prove necessary to apply a post-emergence dressing of nitrogen to maize, this should always be given as a dressing to the base of the plants. It is much better to ensure that an adequate quantity of nitrogen is applied to the seedbed.

The seedbed

Success depends on a very high standard of seedbed preparation; ideally a good frost mould should be obtained by autumn ploughing followed by as few cultivations as possible in the spring to avoid compaction and to conserve soil moisture. The tilth should be level, smooth, fairly fine and able to accept the precision drill. A light rolling across the proposed line of drilling may be necessary to firm the seedbed but the ground must not be so tight that the precision drill has difficulty in achieving adequate penetration.

The tracks, left by the press wheels of the precision drill, should be erased by chain harrowing and the land should then be rolled firm. These operations make it much more difficult for the rooks to find the rows.

CHOICE OF VARIETY

Grain and Silage

It is necessary to match the maturity of the variety to the purpose and situation for which the crop is required. If the variety is excessively early in maturity, the yield is likely to be somewhat lower than it need be but if maturity is too late, growth may be arrested by frost while the crop is still much too immature. In general, the more mature the crop required and the shorter the growing season, the earlier should be the maturity of the variety selected.

Varieties may be classified as early, medium and late, according to their date of maturity. The height of the plants and the potential total yield of dry matter in warm climates increase with lateness but ear development and the date of ripening are correspondingly retarded. Late varieties never reach a suitable stage of development in England and are thus quite unsuitable for English conditions, where only the early and medium maturity groups may be grown.

When maize is grown for grain only, varieties in the early group are suitable (see list below), the precise choice depending on the situation. If the growing season is relatively short, which is the case with much of the potential grain areas, it is safest to keep to the earliest maturing varieties. However, if the situation is really favourable and gives a reasonably long growing season, as in east Devon and low lying parts of Somerset, slightly later varieties (of medium early maturity) are perfectly suitable.

Varieties of Grain and Maize Silage

VARIETIES SUITABLE FOR GRAIN
   (Arranged in order of increasing moisture content at harvest:
            Dekalb 202; Inra 200; Inra 258;
            Pioneer 131; Anjou 196; Anjou 210

VARIETIES SUITABLE FOR SILAGE

EARLY GROUP
Earliest maturity—generally give lower yield; suited to shortest growing season, delayed sowing or where early harvesting is required:
            Caldera 433; Dekalb 202; Inra 200;
            Kelvedon 59A
Medium-early maturity—combine high yields with reasonably early maturity—suited to silage production throughout the south of England.
            Anjou 196; Anjou 210; Asgrow 88; Caldera 535;
            Inra 258; Orla 264; Pioneer 131

MEDIUM GROUP—give a high yield but maturity is rather late and only suited to late harvesting in the most favoured areas in the south of England.
            Asgrow 77; Austria 290; Orla 270;
            Kelvedon 33; United 352

LATE GROUP—Inra 321

Note: The above varieties are all recommended by the National Institute of Agricultural Botany. Selection should only be made after consultation of the recommended lists of maize varietie. 1972, Supplement to Farmers Leaflet No. 2—Green Fodder Crops, obtainable from the Institute.

Whatever the situation, the variety chosen must be early enough to produce mature grain in adverse years. Varieties of medium maturity are invariably quite unsuitable for grain production. When choosing a variety for grain particular attention should be paid to varietal susceptibility to foot or stalk rot; this disease is much more serious with grain crops than with those for silage as the latter are harvested in an earlier stage before the crop becomes seriously lodged.

Earliness is also the most important factor to consider when choosing a variety for *silage*. Ideally the variety chosen should be capable of giving a good yield of practically mature grain in the situation in which it is to be grown. Almost invariably the choice should be made from the early group of varieties which largely complete their growth before the onset of autumn frost. The earliest maturing varieties, which usually give a lower yield of dry matter are suited to marginal areas, for early harvesting or where a silage with a high dry matter content (e.g. made in towers) is necessary. The general purpose silage varieties (of medium early maturity) are suitable for all except the extreme cases mentioned above and give a good yield of dry matter with a high ear content combined with reasonably early maturity.

The 'medium' group of varieties usually give the highest total yield of dry matter but these should only be used in really favoured situations in the south of England where a late harvest is practicable. There is a very considerable danger with this type that frost may stop growth while the ear is still only in the milky stage. The dry matter content of silage from these varieties tends to be lower.

The situation is, of course, occasionally altered by the appearance of individual new varieties, such as the 'early' variety Caldera 535 which outyields the medium varieties. Varieties susceptible to lodging should not be planted on the less sheltered sites.

Sweet Corn

Ear size and yield of ear increases with lateness of maturity. The following selection of varieties, classified in order of picking should prove to be suitable for sweet corn:

(1) Earliking	(6) Indian Summer
(2) North Star	(7) Morning Sun
(3) Golden Beauty	(8) Kelvedon Glory
(4) John Innes Hybrid	(9) Sugar King
(5) Northern Belle	(10) October Gold

The Danger of Saving Seed

It must be emphasised that varieties of maize recommended for grain silage and sweet corn are hybrid varieties and do not 'breed true'. On no account should seed be saved for further planting, as the progeny will be quite different from the original seed purchased. New seed must therefore be purchased for every crop and only named varieties, supplied by seedhouses specialising in hybrid maize, should be planted. Unnamed and uncertified seed should be avoided at all costs; it is dear as a gift.

Plant population

The most suitable plant density for maize depends on the purpose for which the crop is required (Figure 2). With green fodder production, where the only consideration is to produce the largest yield of dry matter regardless of the parts of the plant of which it is composed—yield of dry matter of total shoot—high plant densities give the best results. Plant densities of the order of 14 plants per square yard or about 70,000 plants per acre then appear to be optimal, although there is little critical evidence available for English conditions. At every high seed rates the cost of the seed could be difficult to justify.

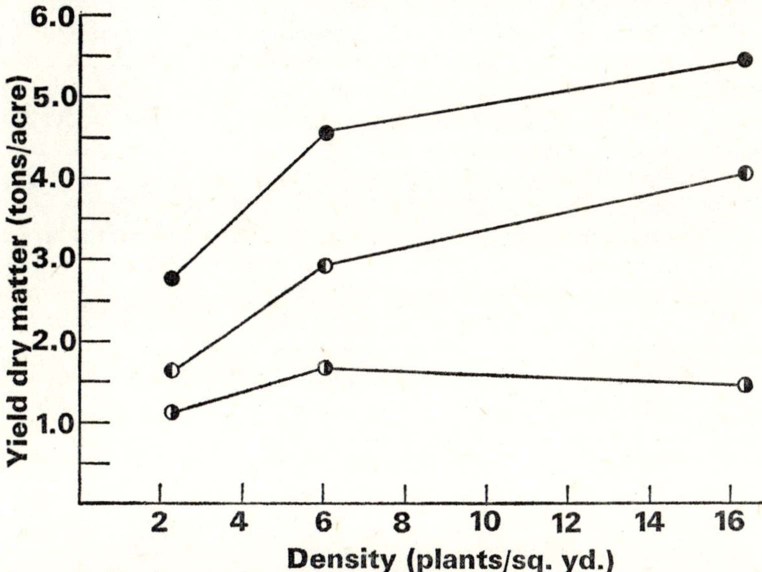

FIG. 2. The Effect of Plant Density on the Yield of Ear and Stover – Maize Variety Wisconsin 275.

Reproduced from *The Cultivation of Maize for Fodder and Ensilage* (Bunting, E. S. and Willey, L. A.)

When the yield of ear is important the situation alters radically. The yield of ear and grain increases with plant density up to the point where any further increase in plant numbers restricts ear production; thereafter the yield of ear declines. With grain production, where nothing but the yield of grain is of the slightest interest, the optimal density is that which gives the greatest yield of ear and grain, i.e. 8–9 plants per square yard or approximately 38,000 to 45,000 plants per acre. The optimum depends on the variety: eg Dekalb 202 needs 38,000; Anjou 210 needs 42,000 plants per acre.

Although a high proportion of ear in the dry matter is of major importance when growing maize for silage it is necessary to obtain an adequate quantity of stover to ensure a high yield of dry matter. The resultant plant population is a compromise, where the optimum, giving an average of one well-developed ear per plant, is slightly above that for grain. The ideal plant population to aim at for silage is thus about 9–10 plants per square yard or 45,000–50,000 plants per acre.

The range which is normally permissible lies between 8 and 11 plants per square yard or 40,000 to 55,000 plants per acre. Too few plants reduces the yield of dry matter: too many results in an excessive reduction of the yield of ear.

With sweet corn only uniform well-filled ears are marketable; undersized or partially filled ears are worthless. Ideally, each plant should produce two good ears, which is achieved by a density of three plants per square yard for late varieties and four plants per square yard for early varieties or 15,000 and 20,000 plants per acre respectively. Too high a plant density restricts ear development and fewer marketable ears will be produced. A lower plant density than that suggested may increase the number of ears per plant but as the plant can hardly fill more than two ears satisfactorily, the outcome is likely to be fewer saleable ears.

Sowing the seed

Provided that the plant population is adequate and that it is evenly distributed within the row, row widths between 18 and 30 inches seem to have little influence on the yield of grain, silage or sweet corn. The row width must be selected to suit the combine, silage-making machinery or the convenience of picking. A row width of 30 inches is normally ideal for grain and silage and most growers appear to be trying to standardise on this row width.

Irregular spacing of the seed in the row is highly undesirable as it reduces the total yield of ear. In some places there are not enough plants, elsewhere bunching of the plants restricts ear production; seed is also wasted. Precision drilling is by far the most efficient way of securing even distribution of the seed and this should most certainly be the normal method of sowing for all regular growers.

If a precision drill is not available or a grower wishes to delay the purchase of a drill until he is satisfied that maize is suited to his conditions, a corn drill may be used. Force feed mechanisms can damage the seed and the proportion of damaged seed delivered by the drill should be checked and the seed rate adjusted accordingly. If available cup feed or pneumatic feed drills are preferable, although it may be difficult to justify the expense of the latter.

The seed should be drilled to a depth of two inches, which is deep enough to ensure adequate coverage; sowing any deeper delays emergence but does not give any additional protection against birds, which usually attack during and just after crop emergence. Slightly deeper sowing may be justified on light soils if it is necessary to place the seed directly on to the soil moisture.

Quantity of seed to sow

Although each parcel of seed is uniformly graded, the size of the seed and hence the number of seeds per pound varies enormously between varieties, between samples and from one year to another within the same variety. This makes absolute nonsense of any flat seed rate recommendation: arbitrary seed rates are thus completely useless. The only satisfactory way of determining the appropriate seed rate is to take the total number of seeds required per acre, as shown in Table 20 and divide this by the number of seeds per pound of the parcel which is actually being sown, as shown in Table 21. The number of seeds per pound usually varies from 1,400 to 1,800.

It is now normal for the merchant to give the number of seeds per pound and the size grade of the seed, thus indicating the size of belt or wheel hole necessary for the particular sample. If not already in stock, appropriate belts or wheels should be ordered well in advance of the sowing date. The seed spacings and seed rates necessary to obtain the required number of seeds per acre at various row widths are shown in Tables 22, 23 and 24.

TABLE 20. Approximate Number of Seeds Required per Acre at Various Plant Densities

No. plants per square yard		Approx. no. of plants per acre (000)	Approx. no. of seeds required allowing for 10% failing to produce established plants (000)
Sweet Corn late varieties	3	15	16·5
Sweet Corn early varieties	4	20	22·0
	5	24	26·4
Grain	6	29	31·9
	7	34	37·4
Grain	8	39	42·9
Silage	9	44	48·4
	10	48	52·8
	11	53	58·3
	12	58	63·8
	13	63	69·3
Green forage	14	68	74·8
	15	73	80·3
	16	78	85·8

TABLE 21. Approximate Seed Rate Required (lb/acre) According to Number of Seeds per lb in the Sample

Number of seeds per lb	Number of seeds required per acre:				
	Sweet Corn		Grain		Silage
	16,500	22,000	40,000	45,000	50,000
1200	14	19	34	38	42
1400	12	16	29	33	36
1600	11	14	25	28	32
1800	10	12	22	25	28

TABLE 22. Approximate Seed Population per acre (to nearest 100) for Grain and Silage Maize

inter row spacing (inches)	row width (inches)					
	20	24	26	28	30	32
$3\frac{1}{2}$	89600	74700	67500	64400	59700	56000
4	75900	65400	60400	56300	52300	49000
$4\frac{1}{2}$	69700	58100	53600	50100	46500	42800
5	62900	52100	48300	45100	41900	39100
$5\frac{1}{2}$	57000	47600	43900	40700	38000	35700
6	52400	43600	40300	37600	34900	32700
7	44800	37900	33800	32200	29900	28000

Note: Row width is largely determined by the harvesting equipment used. Grain-combine/headers are usually set for rows 30 in apart and there seems to be little point in varying this. Two row maize harvesting attachments cannot cope with rows less than 24 in apart.

The required spacing (S) between seeds can be calculated from the following formula:

$$S = \frac{4840 \times 9 \times 144}{\text{row width} \times \text{required plant population}}$$

Reproduced from the *Maize Bulletin* by kind permission of the Maize Development Association.

TABLE 23. Approximate Seed Rate in lb per Acre per Row Width for Grain and Silage Maize

no. of seeds per lb / row width (in)	1200 24–26–28–30	1400 24–26–28–30	1600 24–26–28–30	1800 24–26–28–30
Inter row spacing (in)				
$3\frac{1}{2}$	62–58–54–50	54–50–46–42	46–44–40–38	42–38–36–34
4	54–50–46–44	46–44–40–38	40–38–36–32	36–34–32–30
$4\frac{1}{2}$	48–44–42–38	42–38–36–34	36–34–32–30	32–30–28–26
5	44–40–37–35	37–35–32–30	33–30–28–26	29–27–25–23
6	36–34–31–29	31–29–27–25	27–25–23–22	24–22–21–19
7	31–29–27–25	27–25–23–21	23–22–20–19	21–19–18–17

Reproduced from the *Maize Bulletin* by kind permission of the Maize Development Association.

PLATE 8
The development of swedes and turnips
Photographs 19th November, 1971.

◄ Left to right: swede (Wilhelmsburger—green skin, yellow flesh); swede (Devon Champion—purple skin, yellow flesh) and turnip (Aberdeen Green Top Yellow—green skin, yellow flesh). Note "necks" on the swedes and absence of a "neck" on the turnip.

► Devon Champion Swedes. Left: excessively long neck, typical of close spacing and high nitrogen application. Right: small short neck desirable with swedes for human consumption.

◄ Some shapes of turnip bulb. Left to right: Aberdeen Green Top Yellow (green skin, yellow flesh), Purple Top Mammoth (purple skin, white flesh), Hardy Green Round (green/white skin, white flesh), Con-Continental Stubble Turnip var. "Debra" (purple skin, white flesh).

PLATE 9

The effect of type, variety and treatment on the incidence of plant disease.

The plants in the top and centre photographs were grown on soil heavily infested with clubroot (*Plasmodiophora brassicae*). Photographs, 14th January, 1972, by courtesy of Mr J. L. Jemmett, Regional Trials Officer, NIAB Seale-Hayne.

1. Continental stubble turn —Gelria Tetraploid.
2. Continental stubble turn —Gelria A.
3. Swede—Broadland.
4. Swede—Wilhelmsburger.

Note: immunity of stubb turnips; heavy infection of t swedes. Most of the swed had completely rotted a those shown were among t few left.

5. Rape—Nevin.
6. Kale—Marrow stem.
7. Hybrid kale—Maris Kestrel.

Note: Nevin rape infected but plants still carrying foliage. Marrow stem kale quite unaffected but Maris Kestrel showing some development of clubroot although plants were apparently quite healthy.

Development of powdery mildew. (*Erysiphe polygoni*). on Liragold rape (L_1). Photographed 30th November, 1971. Left: (N_0) plant grown without nitrogen. Note heavy infection of powdery mildew. Right: (N_2) plant grown with 100lb N per acre. Note virtually clean leaves.

PLATE 10

n NIAB trial plot of De Kalb 202 maize at Seale-Hayne Agricultural College, ready for harvesting for
age on 14th October, 1971. Note the highly developed ears of grain.

aas "Jaguar" precision chop forage harvester with single row maize attachment cutting the College
op of De Kalb 202 on 14th October, 1971. Note the absolutely clean pick up by this machine in the
o rows next to the crop; the rest of the rows (ie the headland) were harvested by an unadapted
ail" type forage harvester; the waste, especially of ear, is excessive.

TABLE 24. Seed Spacings Required to Give Appropriate Seed Populations of Sweet Corn

No. of seeds (000 *per acre*) *Row width* (*in*)	*Late Varieties* 16,500	*Early Varieties* 22,000
18–19	$20\frac{1}{2}$	$15\frac{1}{2}$
20–22	$17\frac{1}{2}$	$13\frac{1}{2}$
23–25	$15\frac{1}{2}$	$11\frac{1}{2}$
26–28	14	$10\frac{1}{2}$
29–31	$12\frac{1}{2}$	$9\frac{1}{2}$

Date of sowing

In deciding when to sow, the grower is faced with a dual problem. On the one hand, the sensitivity of the germinating maize seedling to low soil temperatures nullifies any advantage from very early sowing. On the other hand, late sowing allows inadequate time for the ears to reach the necessary state of development before the foliage is desiccated by the first frost of the autumn.

The rate of crop emergence is greatly influenced by soil temperature. Sowing too early in cold soil results in very slow and erratic emergence and is often accompanied by heavy seedling mortality. The plants are then often yellow, weakly and fail to make satisfactory growth; it is therefore only prudent to wait until the soil temperature has reached a satisfactory level. Crops from seed sown when the soil temperature has reached 10°C or above emerge within ten days or so, make vigorous growth and produce large healthy plants, whereas crops sown at a soil temperature of only 7°C may take up to 20 days to emerge and grow more slowly.

Drilling should take place as soon as the mean soil temperature reaches 10°C. Optimum conditions for sowing usually occur during the last few days of April and the first week or so of May. However, if the soil is still cold and wet, which may be the case in a very late spring, it is better to wait a little until the ground is warmer. Sowings for grain should be completed by the end of the first week in May while maize grown for silage should be sown before the middle of the month.

Sowings delayed until late May and early June usually grow rapidly, producing a most impressive looking crop with tall, dark, vigorous foliage but ear production, filling and ripening are seriously retarded. The crop is likely to be far too immature when frost stops growth. Only in the mildest areas, as in the warmest parts of south Devon and Cornwall is such late sowing likely to produce a successful silage crop and even then late sowing is risky and early maturing varieties should be used. Grain crops and silage varieties of medium maturity must never be sown late.

H

Bird damage

Rooks usually constitute the major bird problem and can devastate a promising crop in a matter of hours. Provided the seed is well covered and the drill marks are erased, they do not attack the crop until emergence occurs. They then dig up the seed but discard the shoots, leaving lines of conical holes instead of rows of crop.

The control of rooks can prove to be exceedingly difficult in practice. Provided the acreage is not too large, heavy duty black nylon thread strung on sticks or canes at least three feet from the ground is effective. The canes are placed about 15–20 yards apart in each direction. (Allow 30–40 canes or sticks per acre). The thread should be removed when the maize is 12–18 inches high. It is best to site the maize field as near centres of human activity as possible.

As soon as the crop approaches emergence the field should be visited at dawn, preferably with a gun and any rooks that can be shot should be hung up in the field; they often act as a temporary deterrent. The dawn visits should continue for a fortnight after crop emergence. Automatic guns or strings of crow-scarers are helpful, more so if backed up by live ammunition. Severe rook damage has occurred where the seed has been dressed with special bird repellents but large acreages to which phorate granules have been applied (as a protection against frit fly) in the seed furrow have remained untouched.

Control of diseases and pests

Fortunately maize is not very susceptible to pests and diseases. The seed should always be dressed with a combined fungicidal and insecticidal seed dressing such as thiram with gamma BHC; the operation is usually carried out by the merchant. Maize seedlings are very susceptible to wireworm and leather jacket damage; heavy infestations in poughed grassland should be dealt with by a soil application of a gamma BHC dust or liquid spray, which is worked into the soil during seedbed preparation.

In the west of England grassy stubbles can leave a legacy of leather-jackets for the following year and treatment is sometimes necessary. The crop may be attacked by frit fly, although treatment in a silage crop is usually difficult to justify. In areas where heavy attacks are likely, trouble may be prevented by applying phorate granules in the seed furrow (avoiding contact with the seed) or by scattering chlorfenvinphos granules over the plants when the crop emerges. Instances of severe damage by phorate granules have been recorded in 1971; dosage rate and avoiding contact with the seed appear to be critical.

Growth pattern and harvesting of silage maize

Slight spring frost shortly after emergence is not usually serious and only checks growth temporarily; the plants soon recover as the growing point is well protected. Varieties of European origin are markedly less affected by cold than North American varieties. The initial growth rate of maize appears to be slow until it reaches a height of 9–12 inches in late June. Thereafter, with the arrival of hot weather in July, the crop is capable of making remarkably rapid growth and by the end of the month should be 5–6 feet high with the flower tassels showing. The hotter the weather, the more vigorous the growth; the rate of growth can be most disappointing in cold wet summers.

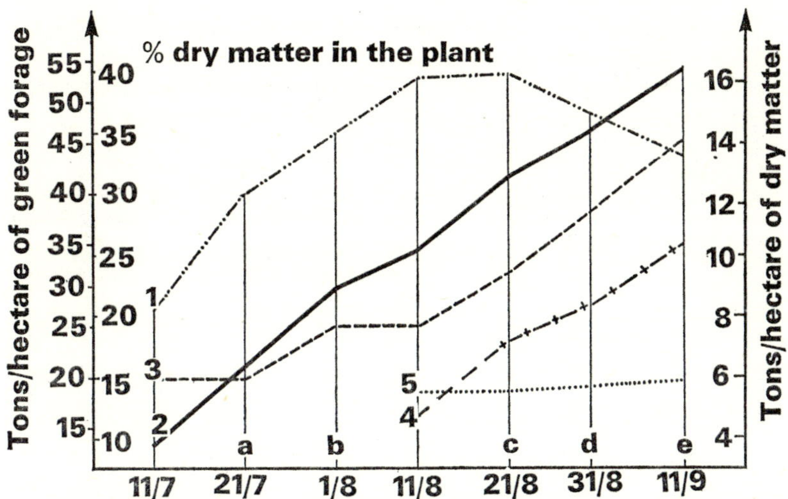

FIG. 3. Development of the Yield of Green Forage and Dry Matter as a Function of Time. (Trials undertaken in France).
(1) *Tons/hectare of green forage.* (2) *Tons/hectare of dry matter produced by the whole plant.* (3) *Dry matter content of the plants (%)* (4) *Tons/hectare of dry matter in the cobs.* (5) *Tons/hectare of dry matter in the leaves and stems.*
(a) *flowering;* (b) *pollination;* (c) *milky stage of grain;* (d) *soft stage of grain;* (e) *hard grain.*

Figure 3 shows that when the crop has only reached the flowering stage, it has produced less than half its potential yield of dry matter and the dry matter content of the whole plant is unlikely to exceed 15 per cent. If ensiled at this stage the crop produces a watery product of very low feeding value and is typical of the so-called 'silage' produced by the very late White Horse Tooth variety many years ago. It is still typical of the result achieved by ensiling maize sown at a high plant density for fodder, very late sown crops or late maturing varieties.

Once the ears have been produced the grains start to accumulate food material, mainly in the form of starch. Thereafter the increase in dry matter is concentrated entirely in the ear, so that by the time the milky stage has been reached the ear contains nearly half the dry matter of the whole plant. At this stage the yield of green material starts to decline but the apparent loss is more than made good by the continued accumulation of dry matter in the ear. Thus the total yield of dry matter, the yield of dry matter in the ears and the dry matter content of the whole plant increase up to the time the crop is ready to cut for silage.

The effect of dry-matter content on the feeding value of maize silage is illustrated by the following two rations for dairy cows; the superior starch equivalent and productivity of the ration made from the silage with the higher dry matter is worth noting.

Dry Matter Content of Silage	Starch Equivalent	Daily Ration	Value of Ration
20%	7	100 lb Maize Silage + 3 to 4 lb Hay	Maintenanee + ½ gallon
25%	13	75 lb Maize Silage only*	Maintenance + 1 gallon

* protein content only adequate for maintenance.

Apart from giving a higher yield of dry matter, a fairly mature crop with a high content of ear—which should make up at least 50 per cent of the total dry matter—is necessary to produce a silage with a high dry matter content and a high energy value. The aim should thus be to obtain an overall dry matter content of 25 per cent when the crop is harvested although this is unlikely to be achieved with varieties of medium maturity and in areas much to the north of the recommended maize area. A dry matter content of 20 per cent must be regarded as the absolute minimum in all circumstances.

The optimum dry matter content varies according to the type of silo. For pits and clamps a dry matter content of 25 per cent or just under is ideal but for tower silos 25–30 per cent is generally preferable. In France, where ripening occurs earlier than in England, it is usually advised that harvesting should be delayed until the crop has reached an overall dry matter content of 30 per cent although in cold years and in late districts a dry matter of 27–28 per cent is the most that is expected.

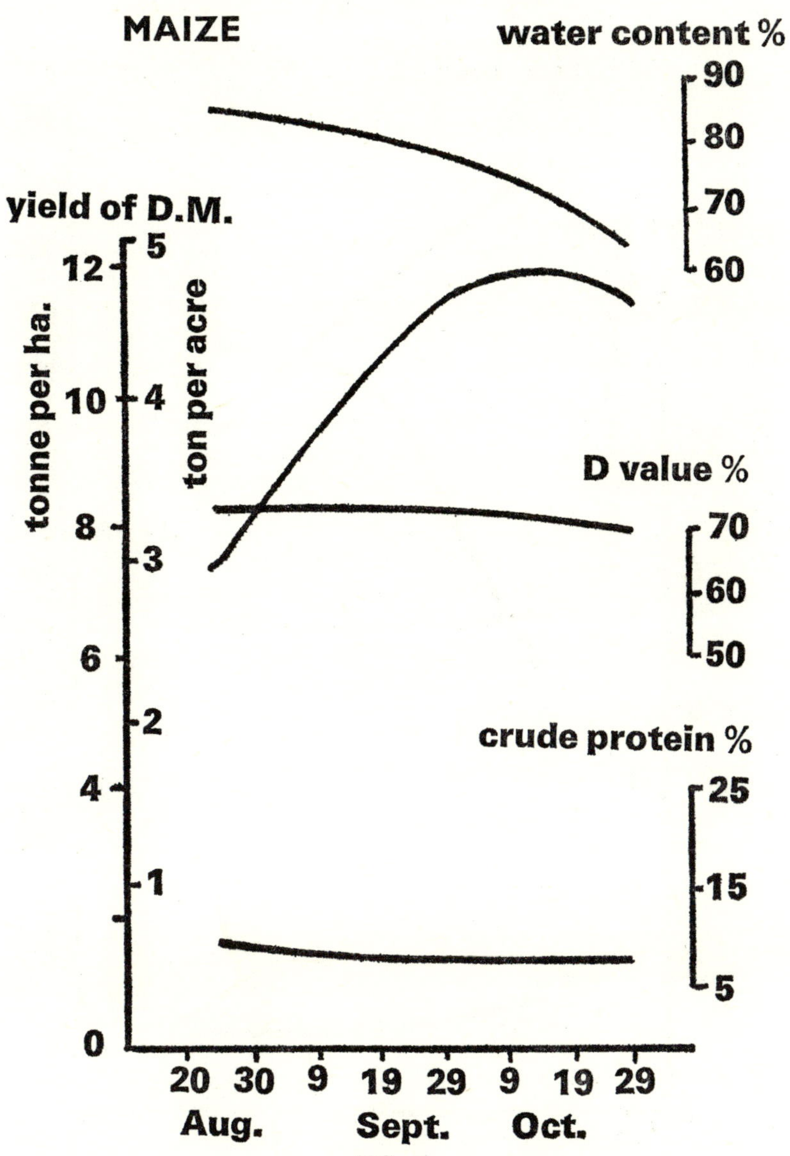

FIG. 4A

FIGS. 4A-D. Changes in Yield of Dry Matter, D-Value, Water and Crude Protein Contents Percentage of (A) Maize; (B) Spring Barley; (C) Winter Wheat and (D) Rye.

Reproduced from unpublished material kindly supplied by A. J. Heard, Grassland Research Institute, Hurley, Maidenhead, Berks.

SPRING BARLEY
water content %
yield of D.M
tonne per ha.
ton per acre
D value %
crude protein %
90
80
70
60
50
40
12
10
8
6
4
2
0
5
3
2
1
70
60
50
25
15
5
10
20
30
10
20
30
9
June
July
Aug

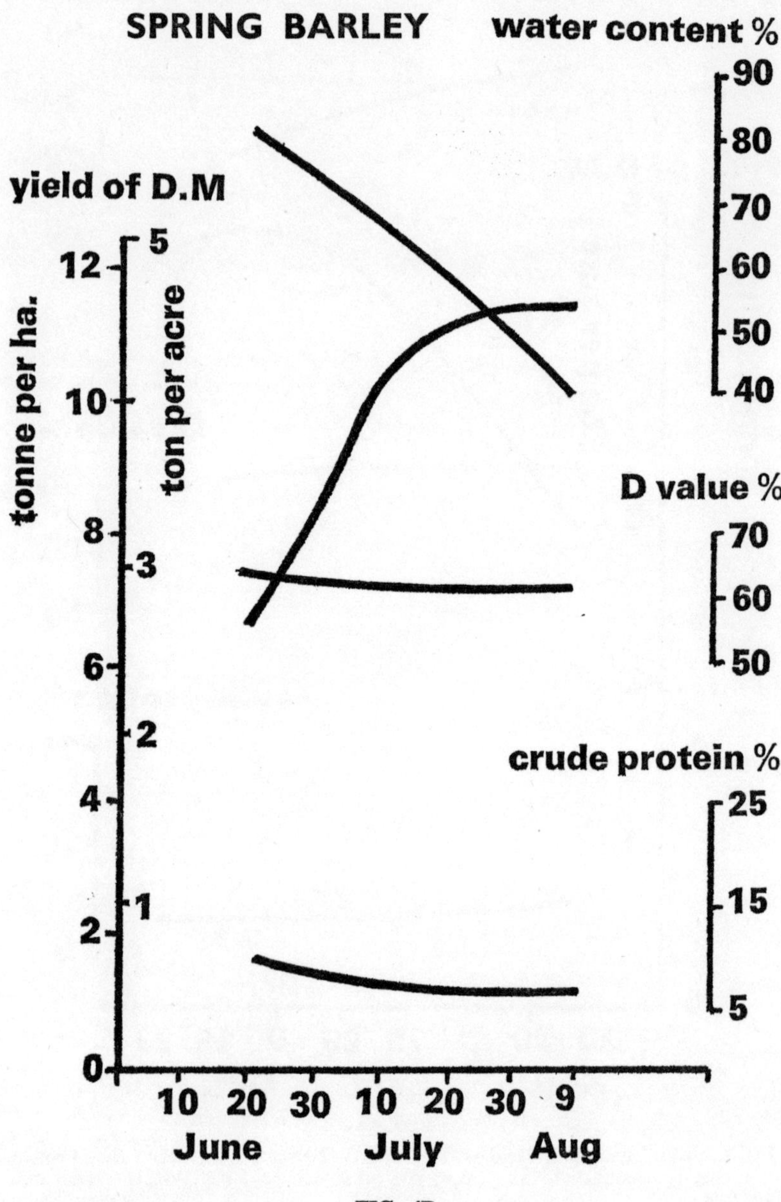

FIG. 4B

WINTER WHEAT
water content %
yield of D.M.
tonne per ha.
ton per acre
D value %
crude protein %
90
80
70
60
5
12
10
4
8
3
6
2
4
1
2
0
70
60
50
25
15
5
10 20 30 10 20 30
June July

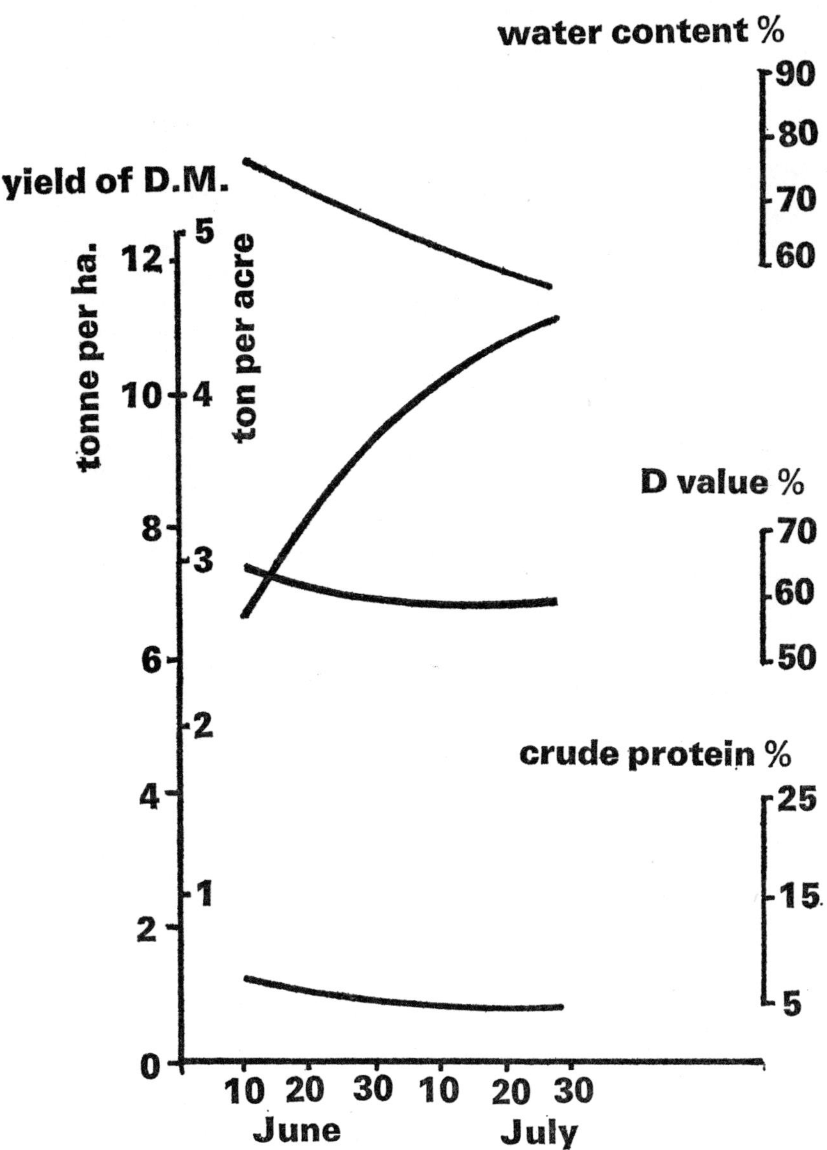

FIG. 4C

RYE
water content %
yield of D.M.
90
80
70
60
tonne per ha.
ton per acre
12
10
8
6
4
2
0
5
4
3
2
1
D value %
70
60
50
crude protein %
25
15
5
10 20 30 9 19 29 9 19
May
June
July

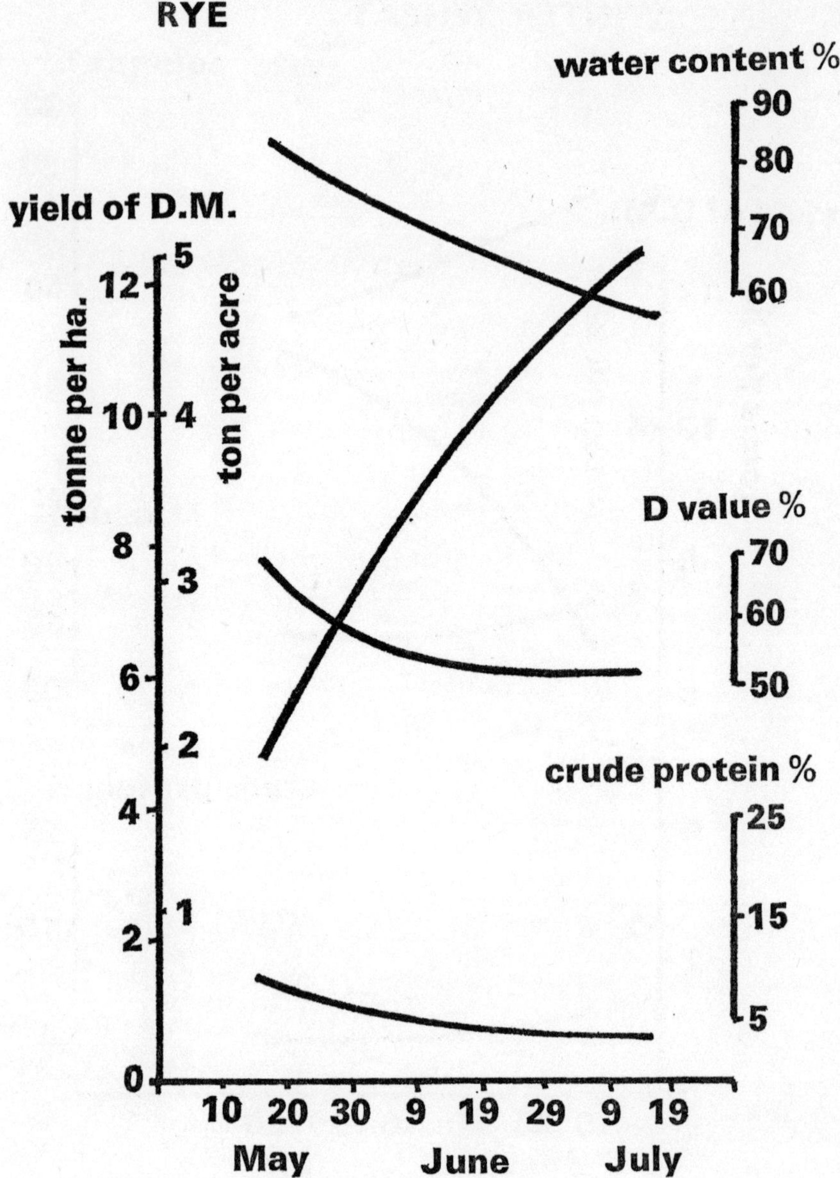

FIG. 4D

The crop is ready for harvest when the grain has attained the consistency of hard dough and is tough to handle. Under English conditions, this stage should be reached in late September or early October (Fig. 4a and Plate 10). Provided the crop is practically ready for harvest desiccation by frost does no harm; in fact it removes some of the water and increases the dry matter content. Once frosting has occurred growth stops and there is nothing to be gained by delaying harvesting any longer. The crop should then be cut without further delay.

The cutting stage for the highest dry matter content (i.e. 30 per cent) is reached when the grains are almost hard and it is difficult to crack them with the finger nail. The husks covering the grain and the leaves at the base of the plant are also beginning to dry out and adopt a papery appearance. There is little change in the digestibility of the plant over a period of several weeks before and during the harvest period (Fig. 4a): the *in vitro* 'D' value of the material should be 68–70 per cent.

Method of harvesting for silage

It is essential to use precision chop harvesters designed or adapted for cutting maize. This type of machine makes a clean job of harvesting and produces a finely chopped material in which all parts of the plant are intimately mixed and which consolidates well in the silo. The result is a silage with a good fermentation, which is easily handled by machinery and which is completely consumed by the stock. Unadapted 'flail' type forage harvesters are quite useless for silage maize. Much of the cob is knocked off the plants and wasted in the field (Plates 10 and 11) while the crop may be badly soiled in wet weather. The length of chop is erratic, thoroughly unsatisfactory, and includes lumps of stem and cob several inches long (Plate 11). This type of material does not consolidate as easily in the silo and is more likely to produce an unsatisfactory fermentation. The silage is also difficult to handle mechanically.

Harvesting and storage of maize for grain

The crop is ready for the combine harvester when the grain is yellow and hard, with a maximum moisture content of 35–45 per cent; this stage is usually reached between mid-October and mid-November. Desiccation of the foliage by frost helps to reduce the moisture content of the grain and enable the crop to pass more easily through the harvesting machinery.

The crop may be harvested by cob pickers or combine harvesters with special attachments but the latter currently appear to be a more practical solution under English conditions. Once harvested the high moisture content necessitates immediate treatment of the grain to prevent heating and consequent ruination of the crop. Drying grain down to 15 or 16 per cent from 35–45 per cent moisture content is usually a slow and expensive operation, requiring two or even three passes over a normal farm drier. A high capacity drier (rating 5 tons/hour) is very desirable. Contract charges for drying down the high moisture content usually experienced with maize in England can often be excessive.

Propionic acid storage would seem to be a much more attractive proposition to the grower who wishes to feed his own grain. This type of storage is extremely efficient; the grain is very palatable, keeps well and dust problems are eliminated. The exact quantity of acid required must be applied evenly and a metered augur-applicator is essential for the purpose; the amount of acid that is needed increases with the moisture content of the grain. Special buildings or sealed stores are unnecessary but it is essential that the building is completely waterproof and that the walls are acid-resistant and capable of standing up to any side pressures that may be exerted by the grain. When storing the grain in a confined space it is necessary to instal extractor fans to remove the fumes during application of the acid if workers have to enter the store to level the grain. The fumes soon disappear and the grain is very palatable to stock and pleasant to handle.

WHOLE CROP CEREAL SILAGE

The basic principles underlying the production of whole crop cereal silage are very similar to those applicable to the production of maize silage. It is necessary to grow a crop which combines a high yield of dry matter with a product having a high dry matter content, a high digestibility and which gives a good type of fermentation in the silo. In both maize and whole crop cereal silage, the immature grain is a vitally important constituent of the silage.

The yield of dry matter obtained from spring barley continues to increase rapidly after ear emergence as the grains develop and begin to ripen (Fig. 4b). The maximum yield of dry matter is reached when the grain is of a mealy or hard cheese consistency and the straw is yellowing. Thereafter the yield of dry matter declines as the leaves fall and some grain begins to shed.

The digestibility of the whole plant follows a somewhat different pattern. Once the stems start to elongate, which occurs early in the life of the plant, the D value of the whole plant declines until the grain develops. The decline then slows down, stops and a constant is reached. This level is maintained for some three weeks in July before the D value again declines as the grain and straw ripen, assuming of course, that they are allowed to do so.

The two main components of the plant, the grain and the straw, behave very differently. As with grass grown for hay or silage, the digestibility of the straw declines throughout the life of the plant; while the digestibility of the grain remains consistently high.

During the period that the weight of digestible material in the grain is increasing, it exactly counterbalances the loss of digestibility that continues to occur in the straw. A constant D value of the whole plant is thus maintained throughout much of July. However, once further accumulation of nutrients in the grain virtually ceases, it is no longer possible to maintain the balance between straw and grain and the digestibility of the whole plant again declines. In some years the digestibility of the plant actually increases during the filling of the grain but this 'lift' in digestibility is erratic in its occurrence.

Suitability of Different Cereal Crops

The behaviour and hence the suitability of the common cereals for whole crop silage differs considerably. Wheat (Fig. 4c) behaves very similarly to barley, but oats, which is the 'traditional' cereal for arable silage production, does not. With oats, the decline in digestibility, which starts with the elongation of the stem, does not show the characteristic early levelling out or the periodic 'lift' produced by spring barley, winter and spring wheat in some years.

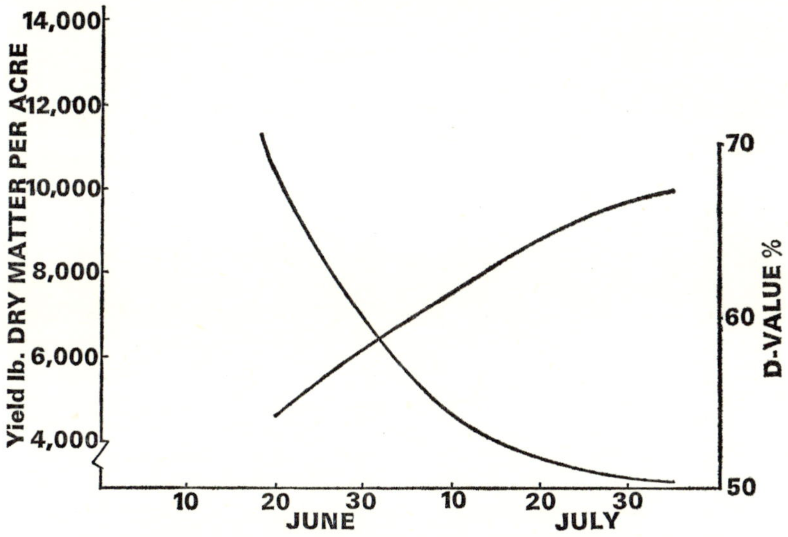

FIG. 5. Dry Matter Yield and Digestibility of Whole Crop Spring Oats (Astor). Harvested in 1966 and given 70 units N fertiliser per acre.

After Raymond, W. F. and Heard, A. J.—*The Ensilage of Whole Crop Cereals. Ceres 1968, No. 4.*

The reason for the continued decline in digestibility of the oat crop (Fig. 5) is probably due to the lower digestibility of the grain. The problem with oats is therefore comparable to the problem met when ensiling grass, although more severe: the period in which a reasonable yield of dry matter can be combined with a satisfactory D value is exceedingly short, as the latter falls very rapidly. Excellent yields of dry matter can certainly be obtained from oats but only at the cost of sacrificing digestibility.

Rye is of very dubious value for whole crop cereal silage, as it has a low D value at any stage of growth at which it is reasonable to make silage and where its yield is comparable to winter wheat or spring barley (Fig. 4d); the two latter crops are much more satisfactory for the purpose.

Winter wheat is capable of giving the heaviest yield of dry matter which is combined with a high D value; cut in mid-July, winter wheat has given five tons of dry matter per acre with a dry matter content of 38 per cent and a D value of 62 per cent in NIAB trials. Although winter wheat is technically very suitable for whole crop cereal silage, it is normally grown as a crop for direct sale and farmers generally prefer to treat it as such. Spring wheat is unreliable for whole crop cereal silage.

Beans, peas and vetches have been extensively added to 'traditional' arable silage mixtures. The protein content of the crop depends on the proportions of cereal and legume. Small quantities of legume have relatively little effect on the protein content. Peas and spring beans usually give a lower yield of dry matter than cereals and the inclusion of these crops in quantity, e.g. in a 1:1 ratio tends to reduce the total yield of dry matter of the crop.

Vetches may give a higher yield than the other legumes but if left too long they tend to become stringy, rot at the base, ruining the quality of the silage; if cut when very immature legumes tend to be excessively watery and can hardly be described as self-wilting.

On balance, barley seems likely to be the most suitable crop and most acceptable to farmers. NIAB trials indicate that a yield of about three tons of dry matter per acre with a D value of 64 per cent may be obtained from spring sown barley, which maintains a good digestibility throughout most of July; results elsewhere indicate that a D value of around 57 per cent is likely to be obtained.

Cultural details

Cultivations are identical to cereals grown for grain. The crop should be sown as early in March as soil conditions permit; late sowing should be avoided as the yield of dry matter is likely to be seriously reduced.

As the aim is to produce a heavy yield of digestible dry matter containing a high proportion of immature grain, lodging must be prevented; high seed rates and excessive nitrogen application should thus be avoided. A maximum seed rate of 112 lb of spring barley should be adequate for practically all situations. The optimum dressing of nitrogen varies according to the soil fertility and is influenced by the previous crop and management, the winter rainfall and the climatic conditions. Trials suggest that the optimum is likely to lie between 60 and 70 units of N although this may require some adjustment according to the level of soil fertility.

Choice of variety

Most short strawed varieties which give a high proportion of grain appear to be perfectly satisfactory. Long strawed varieties are undesirable as their higher straw content lowers the digestibility of the crop. Varieties of spring barley which are unsuited to early sowing or are particularly susceptible to any leaf disease prevalent in the area in which they are to be grown should be avoided. The latter point is especially relevant in the south west of England.

Stage of harvesting

The most suitable stage for harvesting depends on the type of silo to be used. For bunker or clamp silage a rather less mature crop, with a dry matter content of around 30 per cent is preferable, as it consolidates more easily. At this stage, which is reached in early July, the consistency of the grain varies from milky to cheesy. The optimum stage of harvesting for filling tower silos occurs about ten days later, around July 20th, when the moisture content is approximately 40 per cent. The consistency of the grain is by then cheesy or mealy and the straw turning yellow. Any delay beyond this stage is inadvisable as the D value and sugar content of the crop decline. The proportion of grain in the crop and hence the feeding value of the silage may be improved by leaving a stubble some six inches long.

The silages made from whole crop cereals cut at a suitable stage are remarkably uniform and predictable, in fact much more so than the average run of grass silages produced in Great Britain. However, whole crop cereal silage compares less favourably with the best grass silage and is deficient in protein and minerals. Provided the latter deficiencies are made up by concentrate supplementation, whole crop cereal silage may be regarded as a useful bulk food, which is capable of supplying maintenance and the first half to one gallon of milk, according to the stage at which the silage is cut. This silage is well suited to cows at the end of their lactation on which good grass silage would be wasted.

Making maize and whole crop cereal silage

The basic process involved in silage making is the conversion of soluble carbohydrates, such as sugars, in the crop to lactic acid by lactic acid forming bacteria in the absence of air. If conditions are right, the process occurs very quickly and so much acid is produced in the mass of the silage that undesirable putrefying organisms are unable to develop.

Once acidification has taken place, the acid preserves the silage as long as the silo remains undisturbed and the air and water continue to be excluded. However, once the silage is removed from the silo, decomposition can take place and the silage only keeps for a very limited time. In warm weather especially, silage should only be removed from the silo immediately before it is to be fed.

Successful silage making thus depends on the rapid growth of the lactic acid producing bacteria. To achieve this the following basic requirements must be met: the ensiled crop must be rich in soluble carbohydrates; it must have a high dry matter content; and air must be excluded from the fermenting crop in the silo.

The soluble carbohydrates provide the essential raw material from which the lactic acid bacteria manufacture the lactic acid to acidify the mass, while a crop with a high dry matter content precludes the development of putrefying bacteria but allows the lactic acid bacteria to develop, albeit at a somewhat slower rate than in a wetter crop. The exclusion of air is essential as the lactic acid producing bacteria can only operate under anaerobic conditions which also minimise heating of the crop and consequent reduction in digestibility of the silage.

It is therefore hardly surprising that many of the earlier attempts to produce silage from arable crops were dismal failures. The crops were deficient in soluble carbohydrates, especially when vetches which were rich in protein were included. They had too low a dry matter content and the silos of the day failed to exclude air. Foul smelling wet silage, produced by putrefying bacteria was often found at the bottom of the silo while the silage at the top of the silo was often seriously overheated and most indigestible. The latter was also typical of stack silos.

Maize and whole crop cereal silage, cut at a reasonably mature stage, are very rich in soluble carbohydrates, have a high dry matter content and automatically tend to produce the right type of fermentation without additives. Whole crop cereal silage reaches its maximum sugar content when the grain is in the milky stage (circa 30 per cent of the dry matter) after which the sugar content declines. The sugar content is still satisfactory when the grain is in the mealy stage (circa 20 per cent of the dry matter) but drops to about 5 per cent at later maturity as the sugar is converted into starch. Starch does not appear to be converted to lactic acid in the silo. Cutting at such a late stage is therefore inadvisable as the yield and digestibility are declining, the temperature in the silo is less easy to control and an unsatisfactory type of fermentation may occur.

The optimal moisture content depends on the type of silo in use. Relatively mature material with a high dry matter content is more difficult to pack and is inclined to heat rapidly if the air is not completely excluded. Such material is thus only suitable for sealed towers, where the air can be completely excluded.

For filling bunker and clamp silos the crop should be cut in a less mature stage so that it packs more readily. The air can then be easily squeezed out and excluded. Clamp and bunker silos are best filled by the 'sealed wedge' system; once filling has started the work should proceed as rapidly as possible. The stored material must be covered as soon as work stops each night and immediately filling is completed; the top sheet should be well weighted to keep it in intimate contact with the fermenting silage. It is of the utmost importance that there is a really air tight seal between the sheet and the wall of the container or the concrete apron.

Relative values

Maize and whole crop wheat and barley for silage are crops which possess a number of features in common. They are 'self wilting' and can be cut and ensiled in one operation; this eliminates the difficulty of wilting the crop and the losses in dry matter that occur during wilting. The two types of crop are thus suitable for making a high dry matter silage in sealed towers, pits or bunkers; with their high content of soluble carbohydrates they produce an excellent type of fermentation.

They also maintain a fairly high D value over a relatively long period, which gives some greedom in the time at which they may be cut. Both maize and whole crop wheat or barley silage are harvested outside the major grass silage making period in May and June and are thus extremely useful for extending the silage making season; this often allows a farmer to make more silage than he could otherwise manage. The feeding values of well-made maize and whole crop cereal silages are remarkably consistent; however, both are deficient in protein and minerals and supplementation is essential.

The digestibility and feeding value of a good maize silage is considerably higher than that of whole crop cereal silage and compares favourably in terms of energy value with really good grass silage; whole crop cereal silage is only at best equal to the ordinary run of grass silages. Any wheat or barley not required for silage can of course be left for harvesting for grain; this can only be done with maize within the area suitable for grain production and then only from early maturing varieties. A combine with a maize attachment must of course, be available.

The probable roles of maize and whole crop cereal silage depend on the farming system. As wheat and barley are susceptible to take-all, they are useless as break crops and whole crop cereal silage is therefore unlikely to appeal very much to cereal growers. While oats are not affected by take-all they are highly susceptible to cereal cyst eelworm, which may be carried by wheat and barley without either appearing to do so or suffering any apparent damage. The effect of such a 'hidden infestation' can be disastrous; cyst counts on the soil should therefore be obtained from the local advisory service before oats are planted in a run of cereal crops.

In contrast, maize does not suffer from the normal cereal diseases and makes an ideal break crop in the south of England. However, whole crop cereal silage has the advantage where soil conditions make autumn harvesting very difficult and in parts of the country which are too cold or too far north for maize to be reliable. Whole crop cereal silage is also more suited to thin or very stony soils.

It is very difficult to see whole crop cereal silage replacing grass for conservation in the livestock areas of the country; it would seem to be more realistic to regard cereal silage as an extremely useful supplement. In the West Country a number of farmers put a layer of whole crop cereal silage as a sandwich betweeen two cuts of grass silage.

The rotational concepts of growing maize silage and whole crop cereal silage are basically different. Maize is essentially a maincrop where the object is to grow a really heavy yield from the sole crop grown in a season. Once established, the crop grows rapidly throughout the season. On the other hand, whole crop cereal silage only occupies the land for part of the season and a second crop such as kale, rape or Italian ryegrass has to be established.

Under conditions of adequate rainfall the combined production of the two crops can equal that of intensively managed grassland and may form the basis of a very satisfactory system of intensive fodder production. However, the establishment of the second crop can be extremely difficult in dry summers and in low rainfall areas; maize would then seem to be a much safer proposition.

One of the major advantages of whole crop cereal silage is that it makes it possible to obtain a really worthwhile cut of silage from a spring reseed. Furthermore, the establishment of the seeds is often a good deal more satisfactory than under a cereal crop left to mature for grain, as the cover crop is removed earlier, which allows a longer spell of warm weather for establishment before the onset of winter. The seeds are usually undersown and should be sown on a fine firm seedbed immediately after the corn, so that they establish before the silage crop meets in the row.

Immediately the silage crop is removed, the young ley should be grazed over quickly to consolidate the surface and encourage tillering. A dressing of a compound fertiliser, giving 40–45 units each of N, P_2O_5 and K_2O should be applied. The potash should not be skimped, as cereals cut in a 'green' stage deplete the soil of potash much more than cereals left for grain.

With further applications of nitrogen, such a ley can produce very heavily before it enters its first winter, especially if Italian ryegrass is used.

Maize and whole crop cereal silage are also valuable in cases of emergency such as a winter kill of seeds, a sudden increase in stock numbers or when taking over a farm with little useful grassland.

J

Chapter 7

CULTIVATIONS AND SEEDBED PREPARATION

CULTIVATIONS MAY be employed to achieve four distinct purposes:

(i) To drain the soil.

(ii) To prepare a seedbed, disposing of unwanted crop residues where necessary.

(iii) To bulk soil around a crop.

(iv) To control or assist in the control of weeds.

Cultivations undertaken primarily to control weeds are treated separately in the chapter on weed control. Such convenience of treatment is considered to be appropriate, as cultivations are often an essential part of a herbicidal treatment and vice versa.

SOME EFFECTS OF CULTIVATIONS

The function of a 'seedbed' is to provide a suitable soil 'environment' for the germination of the seed and subsequent root development. At the same time the crop must be kept free from competition from weeds and unwanted vegetation until it has formed its own leaf canopy and can suppress weed competition. The effects of cultivation are neither uniform nor necessarily beneficial. Whether or not the intended beneficial result is produced depends inter alia on the conditions under which any given cultivation is undertaken.

For instance, if the soil is too wet, heavy modern tractors produce severe compaction, while ploughs and slipping tractor tyres can smear sections of the soil. The upshot is loss of structure, which impedes drainage and root penetration. On the other hand, cultivations undertaken on a hot dry summer day usually lose valuable soil moisture without which seed is unable to germinate.

If such cultivations are injudicious and severe, such as ploughing which is left to stand for a day and then worked down gradually in the heat, the soil may be dried out practically to the depth of the ploughing and a crop will then be unable to germinate successfully without heavy rain. The necessary amount of rain may not fall for some weeks while a relatively small amount may be just enough to germinate the seed but not enough for further growth; the seed is then 'malted' and dies. Even relatively slight disturbance of the soil surface can encourage the germination of annual weeds which can outgrow the crop seedlings and stifle them. When a fodder crop is to be folded by livestock in the winter, the prior loosening of the soil by cultivations often results in severe poaching.

Direct comparisons between methods and systems of cultivation present considerable difficulty and are often invalid. Apart from inherent natural differences between one situation and another such as soil, weather and season, the quality of judgment and the skill of farmers and their tractor drivers shows marked variation while even the most experienced sometimes make mistakes. Whatever system of cultivation is chosen, a high degree of skill is necessary for its success. It certainly does not follow that either greater depth or amount of cultivation results in any increase in crop yield. The question is often, 'How much cultivation is worthwhile and indeed desirable?'

SOIL STRUCTURE

The pre-requisite to effective cultivation of the soil is the presence of a good soil structure in the top-soil. A good structure is not only a necessity for effective drainage but also for the preparation of satisfactory seedbeds and for adequate root penetration and development. The presence of a good structure is equally essential for any method of crop establishment, whether it is achieved by ploughing followed by conventional cultivations, by minimal cultivations or by direct drilling.

The only satisfactory soil structure is a 'crumb' structure where the particles of the top soil are combined together in small crumbs, granules or 'aggregates', giving the soil a 'friable' feeling when handled. This type of structure is seen in its strongest and most stable state after an old turf or a long ley. Water percolates readily through a soil in this condition and when first ploughed it may be puffy and even a little overdrained, especially for roots and potatoes in the drier parts of the country. It does not therefore invariably follow that the highest yields are obtained from the soils with the strongest structure, although yields from newly broken turf should generally be good.

Adequate amounts of clay particles, organic matter and calcium (i.e. lime) are required for the formation of a satisfactory structure. Soil types therefore differ markedly in the strength of soil structure they can produce. Soils such as sand, which are deficient in clay, do not form a strong structure and considerable quantities of organic matter are necessary to bind the soil together; if organic matter is deficient, sandy soils are loose, lack any cohesion and are particularly liable to 'blowing' when bare ground is exposed in dry spring weather.

The top inch or so of soil, with seed included, is then removed from the field and deposited against the boundary hedges. Lime has no effect on the structure of a sandy soil, although it is highly necessary for crop growth. A pH of 6·0 to 6·5 is adequate.

On the other hand, soils with a reasonable clay content are capable of forming strong soil structures provided they are adequately supplied with humus and lime. In contrast to sandy soils, the pH of heavy clay soils should be maintained at between 6·5 and 7·0. If acidity is allowed to develop, the structure of these soils suffers severely under arable conditions. Maintaining a good structure on clay soils often proves to be quite a difficult problem under arable conditions, although structural problems are by no means confined to clay soils; they can be widespread.

Loss of structure

Loss of structure is usually a slow insidious process. The soil gradually becomes stickier in the winter, is slower to dry out in the spring and requires much more working to obtain a seedbed. As the structure deteriorates, seedbeds become progressively harsher and more difficult to obtain. Eventually the surface drainage may collapse and the soil becomes quite unworkable.

Once the structure has become really bad, the only really satisfactory cure is to sow a ley of at least three years' duration, preferably longer. One year leys are of little value for improving poor structure: they largely maintain the status quo. When improvement of structure is the prime consideration, mown leys are preferable to leys kept very tightly grazed, as the less frequent defoliation allows the grasses to make considerably more root development. Mowing thus results in the production of a greater quantity of organic matter in the form of roots and a stronger soil structure.

In this context, leys for grass seed production have proved to be particularly useful for restoring structure on heavy soils in the Eastern Counties. Severe trouble with structure is less likely on farms where a well managed 3–4 year ley is included in the rotation.

The maintenance of a satisfactory soil structure is largely a matter of good farming practice. Good farming needs the adoption of a sound cropping sequence, returning as much organic matter in the form of crop residues and farmyard manure to the soil as practicable, an especially important point with arable systems, ensuring a satisfactory lime status and avoiding the breakdown of structure caused by mechanical damage from tractors, implements and livestock.

Moisture content

The most important single factor influencing the extent of mechanical damage to the soil structure is the moisture content of the soil, whose physical strength declines as the moisture content rises. Dry soil is relatively strong and difficult to compact but as the moisture content increases up to a certain point, the soil is more readily compacted. Beyond this point the larger pore spaces in the soil become filled with water; although the degree of compaction then decreases, the lack of physical support from the soil results in increased damage from smearing and sinking of tractor wheels, perhaps with the formation of ruts. Loose soil compacts much more readily than soil which has not been previously disturbed. The degree of compaction increases with the weight applied to the soil.

The first essentials are to provide a really efficient system of field drainage and to avoid travelling on the soil or working it when it is wet; this applies equally to grass and arable land. Livestock and tractors can do severe damage to the surface structure of grassland. Not only is grass production likely to be severely reduced in the following year but if it is intended to follow with a direct drilled crop, the roots of that crop will be unable to penetrate the compacted soil and poor stunted growth will result.

It is thus of the utmost importance to keep livestock and the tractors and trailers necessary to feed them off grassland during the winter months. The winter folding of arable crops should also be avoided on wet land.

On the wetter farms, adequate buildings, however cheaply constructed are a necessity. Even temporary yards, unused farm lanes and empty pit silos may be pressed into service on wet farms to keep the stock off the land. However, travelling on wet soil and the resultant poaching cannot always be avoided, especially when roots are harvested in the autumn.

Effect of autumn and winter ploughing

Ploughing during autumn and winter can damage soil structure severely. The level of soil moisture content which gives conditions which are generally considered to be ideal for ploughing coincides approximately with that which gives maximal soil compaction. If the soil is still wetter and thus weaker, grip and wheelslip increases rapidly; damage from smearing and poaching can then be severe.

Although it may be economically feasible on some farms to possess adequate tackle to complete the autumn ploughing of grassland and cereal stubbles while the soil is still relatively dry, this course is not usually possible with root crops and green forage crops used for folding or carting to livestock. A number of fields will almost invariably have to be ploughed later, when the soil is likely to be wet.

Leaving ploughing over until the spring is often inadvisable on the heavier soils, as spring ploughing may result in harsh cloddy seedbeds. The decision on whether to plough as soon as the field is cleared or whether to allow the soil to dry first depends on the soil type, the time of year, and the likelihood of the occurrence of hard frost.

If the soil is fairly heavy and ploughing can be completed by Christmas or early January and is in an area where some hard frost can reasonably be expected, a decision to plough at the earliest opportunity is fully justified.

However, if the season is well advanced, say February or March, the farm is situated in a milder area in the west of England and little hard frost is to be expected, then it is usually much safer to wait until the soil is drier and works satisfactorily.

Like many other decisions in farming, the best course depends on the characteristics of the individual field and is seen most readily in retrospect. Where land has been ploughed in a wet condition, there is much to be said for subsoiling or deep cultivation at the first suitable opportunity.

Soil compaction

It is all too easy to compact the soil during the course of seedbed preparation, especially during the early spring, when the surface tends to dry rapidly but the soil is often still very wet underneath. When several passes are made during the course of seedbed preparation, around 90 per cent of the total area of the field may be rolled by a tractor wheel, and compaction may then be severe. To prevent compaction in the early spring it is therefore necessary to limit the number of cultivations to the absolute minimum and to reduce the degree of compaction at each cultivation as far as possible.

Early ploughing with one-way ploughs generally gives a well-weathered seedbed which needs relatively little working to produce the desired result. Thus when early sown peas or cereals are planted on a well-frosted level tilth, one pass of the drill with seed harrows attached may be the sole cultivation. At most, the drill will be preceded by a single stroke of the harrows or a spring tined cultivator.

The foregoing amount of cultivation may at first sight seem small but the cultivating effect of ridging bodies, drills and seed harrows should not be forgotten.

The seedbed should be in the requisite final stage when all operations are finished—not when the seed is still to be drilled, harrowed in and perhaps rolled.

The most important factor affecting compaction is once again the moisture content of the soil. It is therefore essential to keep off the land until it is fit to work. Patience and the possession of adequate tackle to do the work quickly when the ground is dry enough are the two main necessities.

Mechanical pressure

Mechanical pressure on the soil may be reduced in several ways. Crawler and half-track tractors exert considerably less pressure than wheeled tractors, but they are relatively expensive and only really suitable on large arable farms where soil compaction is a major problem. Cage wheels of the same diameter as the tractor wheels can reduce pressure on the soil considerably. Double wheels and large tyres at lower tyre pressures can also help but at the same time the inflation pressure should be high enough to prevent damage to the tyre itself through underinflation.

When large tractors are used on light work they are often under-loaded, and apart from the resultant lack of economy, they also tend to compress the ground to a greater depth than lighter tractors when tyre pressures of the two are comparable. The effects of compaction by tractor wheels may also be mitigated by growing row crops in 'beds'. On suitable soils conventional cultivations may be replaced by direct drilling.

Effect of implements

Implements used for seedbed preparation either tend to compact the soil, have a neutral effect or may lift the soil. The first group includes rollers, disc harrows and harrows with their tines inclined backwards. These implements are invaluable for squeezing, cutting and breaking clods under suitable conditions but their compacting effect is magnified considerably when the soil is wet, especially that of disc harrows, which are more likely to be used when the soil is unsuitable.

For this reason disc harrows are unpopular with many arable farmers on soils which are at all prone to trouble from compaction.

In the middle group are harrows with vertical tines, which neither compact the soil nor bring clods to the surface. The last group consists of all harrows with forward pointing tines, spring tined harrows and cultivators. Implements in this group lift the soil, loosening it and tending to bring clods and any trash to the surface. The lifting action is particularly valuable in the early stages of seed-bed preparation and also in the early spring, allowing better aeration of the tilth and more rapid drying; however, the tines should not be set so deep that they bring up unweathered soil.

CULTIVATION METHODS

Methods of cultivation may be grouped broadly into three categories:

(i) Deep cultivations for draining and subsoiling.

(ii) Ploughing followed by conventional cultivations.

(iii) 'Minimal cultivations', where ploughing is eliminated and only the minimum number of cultivations necessary to produce a seedbed are given.

(i) Deep Cultivations for Drainage and Subsoiling

If one is prepared to tolerate a low stocking rate with poor financial returns, it is possible to co-exist with indifferent drainage on permanent grassland. However, once a high stocking rate or arable cultivation is introduced, impeded drainage can only lead to disaster. An efficient drainage system must therefore be installed before any attempt is made to plough the land or to grow a crop.

Attempting to grow arable crops on fields with impeded drainage results in the bogging of tractors and implements, the loss or even total collapse of soil structure, delayed cultivations and sowing on bad tilths, poor weakly plant growth and drowning of patches or even destruction or loss of the entire crop. Drainage is therefore the first and most fundamental cultivation that is undertaken.

An efficient system of field drainage is not confined to the installation and maintenance of field drains: the water must be able to pass through the soil to reach the drains and once there, must be able to enter. The ease with which this can occur varies with the soil type.

Provided that the surface has not been previously compacted, surface water percolates readily downwards through a highly permeable soil such as sand, until it reaches the 'water table'. When impeded drainage occurs on such a soil, the problem is normally one of 'bottom' or 'ground' water, when the water comes up from below, or the water table rises too close to the surface. The drainage is impeded simply because the water is unable to escape in any other direction and is typical of low lying sites with insufficient fall. Unless pans or other physical obstructions are present, deep cultivations other than the actual laying of drains have no great effect on the percolation of water in these situations.

Impermeable soils and subsoils

The position is quite different on impermeable soils and subsoils such as clay, where rain and surface water tend to have the utmost difficulty in percolating downwards through the soil. In undrained situations, the water simply lies on the surface or accumulates in hollows and forms ponds. This is a 'top water' problem, in which it is necessary to convey the water from the surface through the upper layers of the soil and into the drains. The whole path of the water must be kept open.

Apart from the need to maintain a good soil structure to allow the free passage of water, it is also necessary to fracture and open up the subsoil, so that the water may reach and enter the drains.

On the heavier soils, where thick uniform clay occurs to within 12 inches of the surface, the normal method of drainage is to draw mole channels over the main drains (clay or plastic pipes); the latter are covered with at least 12 inches of permeable fill, like drainage stone, washed and graded clinker or light industrial aggregate. The moles are drawn across the line of the mains, passing through the permeable fill not less than three inches above the mains, so that the permeable fill allows the water to pass readily from the moles into the mains. Apart from being necessary to allow the passage of the water from the moles, the permeable fill keeps the pipes open and prevents the clay from sealing the slots or junctions in the pipes through which the water must pass to enter the drain.

Except on the more permeable soils, the use of permeable fill should always be considered for covering lengths of pipe whose function is to collect water from the soil. A covering of permeable fill is not necessary for pipes which are merely conveying water down a slope to an outfall. Once a good system of pipes covered with permeable fill is laid, it will last almost indefinitely and only periodic moling is necessary to keep the system fully operative.

The water enters the moles by way of the mole slot which is first reached through the many cracks and fissures produced by the mole plough. Moling is thus not only an operation which lays down cheap lateral or minor drains but is also a most valuable cultivation in its own right, bursting and fissuring the soil and improving the ease of root penetration as well as the drainage.

Mole drainage is still valuable on the less stable clays but the moles do not last as long. The distance between the parallel mains should therefore be reduced to approximately one chain, while the moles are best drawn at a distance of not more than seven feet apart. Mains must, of course, be laid at the bottom of any hollows.

On patchy soils, where the clay is interrupted by seams or pockets of lighter soil, the mole channels are short-lived and may even collapse as soon as they are drawn. The effect is then comparable to subsoiling, which, because it is undertaken in dry soil conditions, exerts a greater shattering effect and may then give more satisfactory results.

Subsoiling

Subsoiling is a very useful means of disintegrating layers of impacted soil, such as plough or iron pans and of achieving very deep cultivation without bringing harsh undesirable subsoil to the surface. With heavy machinery a penetration of up to three feet may be achieved; the deeper the penetration, the wider its lateral effect. The numerous vertical fissures are helpful to both drainage and root penetration. A special subsoiler or subsoiling attachment to a mole plough may be used.

Unlike moling, which has to be undertaken when the clay subsoil is in a reasonably plastic condition in order to form a mole channel with smooth walls, subsoiling is best carried out when the soil is really hard and dry. In this state the soil offers a high degree of resistance to the subsoiler and the maximal amount of bursting and shattering is achieved. A cereal stubble offers an ideal opportunity; the land is clear and at this time of year the soil is usually at its driest, especially after a hot dry summer. The distance between the cuts should be such that the obvious soil disturbance just overlaps that from the previous cut.

The somewhat shallower penetration obtained from really deep stubble cultivation or with bastard fallows, which shatters the soil to the depth of a foot or more in a hot dry summer, is of the utmost value on the heavier loams and clays. A very heavy cultivator and crawler tractor are necessary and this is usually a contractor's job.

The improvements in aeration and water percolation are benefits quite apart from the resultant control of grassy perennial weeds. An identical cultivator is useless on sodden clayland in a wet harvest; it merely cuts deep grooves with practically no soil disturbance.

The heavier soils, then, need good aeration, good drainage and an open structure. Deep cultivation when the soil is dry and natural 'cracking' in hot, dry weather are very beneficial. Conversely, soaking wet summers and the passaging of combines, tractors, heavy implements and livestock on wet soil pack it down, stifle it and make drainage and subsequent cultivations matters of no small difficulty.

(ii) Ploughing Followed by Conventional Cultivations

Before the introduction of herbicidal techniques to control grass and grassy weeds, ploughing was the only satisfactory method of disposing of unwanted surface vegetation and of providing a clean surface on which to produce a seedbed. The effect is often relatively short-lived and regeneration of the buried grasses occurs frequently, especially where turf is inadequately buried. The burial of the storage organs of perennial weeds merely serves to ensure their survival and subsequent propagation; they soon grow again.

The mouldboard plough is traditionally the first step in seedbed preparation but its value has been increasingly questioned in recent years. Ploughing consumes a considerable amount of time and power, it can bring stones and unwanted subsoil to the surface and damage soil structure. The necessity and in fact the desirability of ploughing depends on the crop and on the circumstances.

Crops which do not require a deep seedbed, such as cereals, grasses and clovers and a number of the brassicas, can be grown perfectly well without ploughing. Provided that weeds and unwanted grasses are adequately controlled and that the final seedbed produces satisfactory growth, the precise method by which a seedbed is prepared for these crops seems to matter little. Ploughing, minimal cultivations and direct drilling have all given satisfactory results under suitable soil conditions.

The situation is different with crops such as sugar beet and potatoes, whose harvested parts develop below the soil surface. Experimental results have then shown that the yield of these crops can be seriously reduced by the omission of ploughing. The use of the plough is also usually necessary for growing roots on the ridge.

The appropriate depth for ploughing depends on the crop and the soil. With cereals, grass and brassica crops for fodder, ploughing at a depth of over five to six inches is usually quite unnecessary; only perhaps with crops grown on the ridge, is deeper ploughing generally worthwhile and then only when there is a good depth of rich top soil. Should deeper ploughing than usual be undertaken, not more than half an inch of subsoil should be brought up at any one time and then only in autumn or early winter, when it can reasonably be expected that there will be enough hard frost to weather the exposed subsoil adequately.

In the past, potentially useful soils have been utterly ruined by bringing up large amounts of raw unweathered subsoil in a single operation. Most subsoils are best left where they are and subsoiling is usually preferable to deep ploughing. The response to deep ploughing has often been very disappointing.

In Scotland and the North of England, roots are traditionally grown on the ridge, notably in the higher rainfall areas and on the stiffer soils. In these circumstances ridge culture has a number of advantages. Not only is surface drainage improved but the roots sit well on top of the ridge and remain much cleaner, especially when they are folded by sheep. Conventional methods of singling and weed control are also easier than on the flat. Growing roots on the ridge is undesirable on the lighter and drier soils, as the ridges are rather prone to drying out. Ridge culture is unsatisfactory with mechanical harvesting.

(iii) Minimal Cultivations

Provided that the ground is free from excessive loose trash and unwanted surface vegetation is effectively controlled by the application of an appropriate herbicidal treatment, minimal cultivations possess a number of advantages over traditional methods of cultivation with the plough; this is true both for the control of grassy perennial weeds in cereal stubbles and for the preparation of seedbeds for crops which do not require a particularly deep seedbed, when elimination of ploughing reduces the labour and power requirement considerably.

In some instances, such as sowing grass seed in cereal stubbles, substantial cultivations are in fact undesirable and all that is necessary is to produce a shallow surface tilth adequate to cover the seed. Here, ploughing normally wastes time, delays sowing and loosens the soil to a depth of several inches so that considerable work is necessary to return the seedbed to an adequate state of consolidation; where trash occurs the stubble may be burned or raked up before working starts. Ploughing should be the last resort.

Minimal cultivations are especially applicable to very shallow soils, where ploughing is almost certain to bring up unwanted subsoil. This problem may occur quite frequently and examples range from the reseeding of old worn out grassland where there is only a thin layer of topsoil overlying thick intractable clay to cereal growing on the thinner chalk and limestone soils. By avoiding ploughing, the organic matter from the previous crop or turf is retained in the top few inches of the soil, where it helps to bind the soil together and exerts a strong stabilising influence.

This is especially the case on some of the lighter sands, which are subject to blowing, although here it would seem to be logical to refrain from disturbing the soil surface wherever possible and to resort to direct drilling once the surface vegetation has been reduced to a suitable condition.

When old grassland is to be reseeded straight back to grass or broken for arable cultivation it may need considerable preparation first. The surface may be seriously compacted or there may be a very substantial amount of trash or even a mat of undecayed organic matter present. The application of lime, slag, herbicides to kill the broadleaved weeds, better grazing management and the use of ripper harrows a year or two before operations begin can work wonders:

so much so that the result may dissuade the owner from breaking the pasture at all.

Assuming that seed is to be established, destruction and disintegration of the turf and mat are first necessary and the compacted soil must be shattered; cultivations are usually necessary. For reseeding with grass, desiccation of the sward by paraquat followed by minimal cultivations, when a rotary cultivator is ideal, are usually the least that is necessary.

There is often much to be said for delaying reseeding until after a break of one or more arable crops has been taken. Apart from cashing in on the stored fertility this course gives clean land, a good seedbed for sowing the grass and may well eliminate the danger of carrying over some animal parasites and diseases. When matted old grass is to be first brought into arable production, desication of the sward with paraquat followed by thorough pre-ploughing cultivations is usually well worthwhile. A bastard fallow is excellent in a hot dry summer.

Ploughing the old sward down without any previous cultivation or application of herbicide and then planting a crop at once is usually thoroughly unsatisfactory and may even prove to be disastrous. Once deeply buried, a mat takes a long time to rot; it must be broken up before ploughing. The resultant seedbed may also be so spongy or springy that the crop is unable to establish, while much of the turf may regenerate, reducing yield and causing considerable trouble with grain crops at harvest time.

Seedbed Requirements

The differences between crops in their seedbed requirements are generally widely appreciated. The seedbed conditions most suited to each crop or group of crops may be outlined as follows—

(a) Root and brassica forage crops: When crops with small seeds are grown on seedbeds prepared by conventional means, the condition of the seedbed is critical; the seeds can neither germinate nor develop satisfactorily in rough dry cloddy conditions. Evenly developed vigorous stands can only be established where the seed has friable but moist soil pressed around it. In the days when root crops were sown relatively thickly and singled manually, some irregularity in the stand was relatively unimportant; experienced singlers could usually make good any deficiencies by careful selection of the plants which they left.

Now that the seed of most root crops is sown thinly with a precision drill, any substantial proportion of misses becomes a serious matter. It is quite impossible to drill either accurately or at an even depth on rough or knobbly tilths. Residual or soil applied herbicides are also quite useless in such circumstances. Really high quality seedbeds with a fine firm level surface are therefore an absolute necessity where precision drills and pre-emergence residual herbicides are to be used.

(b) Maize: Although maize has a large seed, it is highly sensitive to poor tilths and adverse soil conditions. A fine level friable tilth is essential for satisfactory growth and for precision drilling this crop. With poor cloddy seedbeds, especially if the weather turns wet and cold, maize seed either rots or produces stunted little yellow plants which make no growth whatever.

(c) Italian ryegrass: Again a fine firm seedbed is necessary, although depth is unimportant provided that there is enough soil to cover the seed. Relatively shallow seedbeds are often preferable as they are more readily consolidated.

(d) Cereals (barley, oats, rye and wheat) and the large seeded fodder legumes (beans, peas and vetches): With autumn sowing, a fine surface is undesirable, especially on a soil where the surface is inclined to run together and form a cap or crust through which water has difficulty in penetrating. Wherever possible it is best to leave a moderately rough surface, although the clods should not be excessively large; at the same time there should be adequate crumb to receive the seed. Puffiness is particularly undesirable but is mainly a problem when grassland, especially an old turf, is first ploughed; thorough consolidation of the seedbed is then of the utmost importance.

The seedbed requirements for spring sown cereals depend on the date of planting and on whether the cereal is to be undersown with grass or clover seeds. For early sowing, say mid-March or earlier, the soil beneath the surface is inclined to be wet and over-consolidation can readily occur. Working should therefore be minimal and a very fine seedbed should be avoided, especially in soils inclined to run together if the weather turns wet; some clod on the surface helps to prevent capping. It is usually inadvisable to leave the soil rolled except perhaps in the case of the lighter soils and in drier situations.

With later planting—late planting is not to be recommended for high yields of cereal silage—and when undersowing with grass and clover seeds, fine firm seedbeds are essential and the roller must follow the drill. Really good consolidation is especially important with late sowing under dry conditions; a loose puffy seedbed will then almost certainly lead to complete disaster.

Moisture Conservation and Crop Establishment

The problems likely to be encountered during seedbed preparation and crop establishment vary according to the time of year. In Great

Britain the annual 'precipitation' (rainfall plus snow) is unevenly distributed. About two-thirds falls in the six 'winter' months (October to March) while only one third falls in the corresponding 'summer' period (April to September), which constitutes most of the effective growing season.

Throughout most of the 'winter' period precipitation exceeds the small amount of water lost from the soil by evaporation. The land is therefore generally wet and the basic cultural problem is to keep the soil dry enough for cultivations to start as early as possible in the spring and to extend as late as possible into the autumn. Plant growth is rarely restricted by lack of moisture in the winter; the problem is more often to prevent the root system from being restricted or even killed by water logging; in short, it is necessary to provide an efficient drainage system. The factor limiting growth in the winter is normally low temperature.

In the 'summer' period the position is reversed. The amount of moisture lost from the soil by evaporation normally exceeds that returned as rainfall, so that with a basic situation of soil moisture deficiency, there is often inadequate moisture in the soil for the germination and initial development of summer sown crops. Moisture is lost from the soil either by transpiration through the leaves of plants, a process which is necessary to cool the leaves and maintain growth, or from the soil surface, especially during the process of cultivation and immediately afterwards.

In neither case is the loss constant. Once the surface inch or so of the soil has dried out, the loss of moisture from bare ground which is kept free from living vegetation is practically negligible.

Loss by transpiration

When a crop is sown, the moisture lost by transpiration of the young plants is initially small but rises as the plants increase in size. Once the leaves of the plants meet and form a complete canopy, the crop is capable of transpiring the maximum amount of moisture; this is the case with all vegetation, whether it is forest, grassland, arable crops or weeds. However, as the crop matures and the leaves die back and drop, the rate of transpiration falls sharply and stops altogether.

Thus when grass, clover or a green crop such as cereals, beans or vetches are taken for silage in dry weather, they may have withdrawn so much moisture from the soil that there is not enough moisture left for the germination of the succeeding crop. Conversely, once a crop of maturing cereals has ceased to transpire and rain falls before harvest, the standing crop tends to shield the soil from the direct rays of the sun and to some extent act as a mulch, tending to reduce loss of moisture. If a second crop is drilled immediately after removal of the cereal crop and the accumulated moisture is not dissipated by cultivation, germination can occur at once and establishment is rapid. A sward desiccated by herbicidal treatment behaves in a similar manner to a mature cereal crop; transpiration ceases with the death of the leaves, which again act as a mulch and prevent the loss of further soil moisture.

Influence of cultivations

Cultivations exert a very large influence on the moisture status of the soil, which they can either help to accumulate, to preserve or to dissipate. Techniques aimed at the accumulation of soil moisture are not new; the value of the fallow for moisture conservation has long been known in the hotter countries. Two variations of the fallow technique by which soil moisture may be accumulated are relevant to the present discussion.

Timely autumn ploughing does not merely open up the soil for weathering by the frost, but also allows the soil to absorb maximal moisture from the winter rains. The spring or summer variation of the technique is applicable where a second crop follows the first in the same season. The choice then lies between either breaking the ground immediately after the removal of the first crop and preparing the seedbed for the second crop in gradual stages or in attempting to intensify the system as much as possible and sow the second crop immediately after the first.

At first sight an immediate turn-round between crops would appear to be the best solution, especially from the economic point of view. This course is ideal where there is ample soil moisture and the soil surface is not disturbed, as with direct drilling. The second crop is usually a quick growing low cost fodder crop such as rape or kale.

Considerable difficulties can arise when these pre-conditions are not met. If dry weather prevails, which is a strong probability in areas with a low summer rainfall, the first crop is likely to have removed so much of the soil moisture that little remains for the second crop; still further moisture may well be lost by ploughing and subsequent cultivations.

In any event, these 'instant' methods of attempting to produce a seedbed by a series of cultivations often tend to result in a harsh unkind tilth which is unsuitable for small seeds, especially on soils with a relatively high clay content.

Sowing immediately the ground is worked also leads to severe problems with annual weeds, as there is insufficient time to germinate and destroy the weeds before the crop is sown. Herbicides may of course be used, but in some crops, particularly the brassicas, the number of suitable herbicides is limited and these are mostly effective only against a very limited range of weeds.

The decision on whether to use a 'fallow' or 'stale-seedbed' technique to produce the seedbed gradually is determined in no small measure by the value and importance of the second crop. While it is feasible to take some risks with the second crop in situations where a heavy yield from the first crop is a matter of prime importance, as with a cut of silage on a dairy farm, the risk of failure must be reduced to minimal proportions when the second crop has to take precedence; the latter situation is typical when the second crop is a high value cash crop such as brussels sprouts or winter cauliflower or a main crop of swedes.

It is then necessary to limit the life and production of the first crop in order to conserve moisture and to allow adequate time to prepare a suitable seedbed for the second crop. The first crop may thus be rye for early grazing or perhaps Italian ryegrass from which only an early bite is taken; the temptation to obtain a further cut must be resisted at all costs.

As soon as the first crop is removed the ground is broken, worked to a partial tilth immediately and the seedbed is prepared gradually with periodic harrowing. This treatment usually results in fine friable tilths, conserves rainfall so that there is ample moisture to start the crop and gives a very useful control of annual weeds provided that rain falls during the course of operations. The 'stale seedbed technique', involving the use of a contact type of herbicide may be used in conjunction with this process. (Chapter 9.)

The timing and manner in which cultivations are carried out has a large effect on the extent of the moisture loss incurred. Under summer conditions losses of moisture are at their maximum on a hot day with a dry atmosphere, especially when there is a warm breeze as in good haymaking weather. Any ground disturbance under these conditions therefore leads to the most rapid loss of moisture and is best avoided.

Conversely, evaporation is slight or may even be non-existent under cool humid conditions, as in misty or cloudy weather and in the early morning and late evening. When there is a deficiency of soil moisture there is much to be said for restricting cultivations to days and parts of the day where moisture losses are likely to be minimal.

Seedbed preparation for immediate sowing

When it is necessary to prepare a seedbed during the summer for immediate sowing, speed of completion of all operations becomes vitally important; in fact speed can make all the difference between success and failure. The shorter the duration of the exposure of

K

moist soil to sun and wind, the less moisture is lost. Even a single passage of an implement can dissipate a large amount of moisture on a very hot day.

When ploughing is necessary to bring moisture to the surface to germinate the seed, the best approach is to plough, cultivate, sow and consolidate the land in relatively small sections; the ideal area of the section depends on the size of the tractor force and should not be more than can be conveniently dealt with in three or four hours.

A particularly good arrangement, which has worked well in hot weather, is to undertake all operations during the coolest part of the day like the late evening and early morning. Ploughing and rolling may start in the late evening or at dawn and the final rolling should be completed before the dew is finally dispersed, say by nine or ten o'clock in the morning. Brassicas planted in this manner have been known to emerge in four days—and that during a drought.

General Approach to Moisture Conservation

Apart from providing irrigation facilities, which is often impracticable and rather expensive for fodder crops, much can be done to improve the supply of soil moisture in the summer. The susceptibility of a crop to drought is determined to a considerable extent by the moisture holding capacity of the soil and the effective rooting depth of the crop. A deep tilth layer with an ample supply of organic matter and a good structure, which can hold a maximal reserve of moisture and through which roots can make deep penetration into a well fractured subsoil is most helpful.

The more organic matter that can be returned to the soil by way of crop and animal residues, the better. Provided they have become well incorporated into the soil before the advent of dry weather, generous application of nitrogenous fertilisers is particularly helpful, especially with crops such as kale; to some extent nitrogen and water are interchangeable. Early sowing in dry areas ensures that there is enough moisture to start the seed and for the crops roots to achieve adequate penetration before really dry weather sets in.

Seedbed preparation by conventional means

Such is the wide range of soil and climatic conditions occurring in British farming that lack of space precludes any comprehensive or detailed discussion of the practical details of seedbed preparation; treatment must therefore be limited to a brief summary of the more salient features.

The techniques broadly applicable to the different types of seedbed required may be conveniently grouped according to the time of year that ploughing (or other suitable method of primary cultivations) occurs.

Autumn and Winter Ploughing

(a) *Autumn-sown cereals*

Cereals which are likely to be required for grazing or fodder production should be sown reasonably early, which ensures that they are sown and established before the onset of wet weather and difficult soil conditions. Early completion of ploughing allows settling to occur. The soil should be adequately consolidated but not compacted, especially after turf. While in the furrow, the soil drains reasonably well but once the furrow is broken, the soil retains much more water and the land may be slow to dry out.

It is, therefore, usually inadvisable to work down too much ground ahead of the drill. In a wet autumn, land should only be worked as it is wanted for drilling and the number of passes should be minimal. Fine autumn seedbeds should be avoided where possible.

(b) *Autumn ploughing for spring crops*

Wherever practicable, stubble cleaning should be undertaken before autumn ploughing for both spring corn and roots, both in the interests of stubble hygiene and weed control. The autumn is also a convenient time to apply dung for root crops; the dung is best built into a heap at the side of the field as it is moved from the yards and buildings during the previous winter and spring.

Wherever practicable, ploughing is best completed by mid-December, especially on the heavier soils in the colder areas. Clay soils are notoriously difficult to work and the production of a fine seedbed for small seeds in the spring is well-nigh impossible without satisfactory frost action. These heavier soils do not respond to late ploughing followed by an attempt to force a quick seedbed.

However, in the milder south west, early completion to the autumn ploughing may result in some very green ploughland in the following spring; spraying with a herbicide is then necessary. In livestock areas, typical of much of the south west of England there is a good case for keeping a considerable part of the land covered by green crops and short leys during the winter to obtain maximum fodder.

With spring cereals grown for fodder early sowing is necessary to secure maximal yields of green material. Patently the land must not be worked before it is fit and it is usually best to allow a time lapse of a few hours after the first pass of the harrows for the disturbed soil to dry to a friable state before proceeding with further cultivations.

Cultivations should be kept to the minimum consistent with the production of a satisfactory seedbed. If difficulty is experienced, which may be the case on heavier land after a mild winter, it is best to accept the best seedbed that can be obtained with limited cultivation and sow the crop on time. Trying to obtain a finer seedbed in these circumstances usually results in severe soil compaction and delayed sowing of the crop.

With spring sown root crops such as mangolds a 'stale' seedbed produced by autumn ploughing is the most effective. The first cultivation at least is best done by a tined implement, further working becomes increasingly shallow and care should be taken not to bring unweathered soil to the surface. Where time and rainfall permits, gradual working allows annual weeds to be germinated and killed during the course of seedbed preparation. Apart from the occasional rolling to break down clods there is no hurry to firm the seedbed until the final stages.

The ultimate 'fining' of the seedbed is best achieved by alternating the harrows and a cambridge roller. On no account should a fairly fine seedbed which remains unsown be left rolled; if heavy rain follows, all but the lightest soils tend to run together and become compacted. The seedbed is then lost.

Spring and Summer Ploughing

Whether sowing is to be carried out immediately a seedbed can be prepared or a 'stale seedbed' technique is to be used, the land should be ploughed and worked quickly to at least a partial seedbed to conserve moisture. A digger or semi-digger plough, set to throw as flat and as broken a harrow slice as possible is ideal. A roller normally follows the plough closely.

When the crop is to be sown at once, the land is worked to a seedbed as quickly as possible, drilled and finally rolled down tightly. The heavier soils are not usually suited to this technique; in the absence of frost action they need a period of baking in the sun or alternate wetting and drying to break down the clods.

The 'fallow' or 'stale seedbed' technique broadly follows the same principles as that already outlined for the sowings made after winter ploughing except that considerably more working is often necessary. When the land is being prepared for crops such as cabbage and brussels sprouts thorough working of the soil is especially important and an adequate depth of moist friable soil should be prepared to take the transplants. The field should always be waiting for the plants: the position should never be the other way round.

Sowing the Seed

The introduction of the precision drill, which sows seed singly and at pre-determined intervals has transformed the labour situation with root crops. Instead of the laborious job of singling which was necessary with the traditional thick sowing, the need for hand work has been reduced to chopping out the unwanted plants or even eliminated altogether. By far the greatest part of the acreage of roots grown for fodder is now precision drilled; the precision drill is also particularly suitable for maize, as it is necessary to achieve a fairly exact and evenly distributed plant population.

The earlier method of establishing root crops with the precision drill was to space the seed some 1–2 inches apart and remove the unwanted plants by hand. This approach is still the safest, especially under less favourable conditions and least likely to lead to a plant population being too low or too unevenly distributed. The labour requirement may be further reduced by the use of a wider spacing, say three inches between the seed but the risk of there being too few plants if conditions are unfavourable is correspondingly increased. A more recent development is the technique of drilling a 'stand' where only enough seed is drilled to achieve the final plant; seed is then usually spaced four or even six inches apart, according to row width and singling is eliminated.

The wider the spacing, the greater the risk and the more dependent one becomes on near-perfect seedbed and weather conditions and the absence of pests such as pigeons. With sugar beet, for instance, drilling to a stand has proved to be reasonably successful in the moister western parts of the county but there have been some most unfortunate results at times in the drier eastern counties.

The most suitable approach

The most suitable approach depends on the value of the crop, the size of root required and the labour available. If a small acreage of mangolds is being grown, a really heavy crop is required and labour can be spared to chop out the unwanted plants, it would seem logical to space the seed two to three inches apart to ensure an adequate final plant population. This approach is widely adopted with swedes grown for feeding beef cattle in Scotland and the north, where a really heavy yield of large roots is a prime consideration.

Removing unwanted plants takes only a fraction of the time taken to single a traditional stand and the plants can be left rather longer without adverse effects. The timing of singling is therefore less critical and a smaller labour force can deal with a larger area. On the other hand, drilling swedes to a stand has become commercial practice in Wales and the south west of England where smaller swedes are grown for sheep feed and also, in the latter case for market. Interest in this technique is widespread and could well increase still further as herbicidal techniques improve and labour becomes even scarcer.

To obtain an adequate and reasonably distributed plant population it is necessary that a very high proportion of the seed should establish, especially when drilling to a stand. A fine level weed free seedbed is the first essential and the seed must be dressed with a suitable fungicidal/insecticidal seed dressing; if brassicas are not so protected against flea beetle, a gappy plant or even loss of the entire crop is more than likely.

Only properly graded seed should be used and the appropriate belt or wheel must be available on the farm to suit the grade of seed purchased. This check must be made well in advance of the intended sowing date. With kale, turnips and swedes graded seed is used. In the case of fodder beet and mangolds either rubbed and graded seed or pelleted seed is obtainable. The approximate seedrates obtained at various seed spacings and row widths are shown in Table 25. The size of maize seed differs greatly from one variety to another and it is most important to obtain advice from the manufacturer of the drill as to which belt or wheel to use for a given maize variety (Chapter 6).

As with cereals and grass/clover mixtures, rape and turnip seed are frequently broadcast when grown as catch crops; kale is sometimes treated in this manner. Broadcasting is a quick operation and perfectly satisfactory under moist conditions but is very much less certain in dry weather and in dry situations. Drilling places the seed positively in the soil and provided there is adequate moisture the time taken for germination is minimal. Drilling is considerably more economical of seed, an extremely valuable point when the seed of some types costs as much as £1.00 to £1.50 per pound.

When drilling replaces broadcasting, a normal corn drill with a 7-in coulter spacing is satisfactory; all coulters should be used for catch crops. The seed may require bulking to ensure that an economical seed rate is sown through the seed box without mincing the seed in the feed mechanism, as most corn drills are not designed for sowing small seed. Suitable diluents and the method of drill calibration are discussed in Chapter 8.

With combine drills the seed may be mixed with granular superphosphate and sown through the manure box at rates of up to 1 cwt per acre. It can be dangerous to exceed this amount, especially with concentrated compound fertilisers. Seed and manure may be conveniently mixed with a small concrete mixer: seed quantities must be weighed carefully.

Unless using precision drills, which have their own press wheels, it is important to drill on a firm surface; it is often a good practice to put a roller in front of the drill; the roller usually follows the drill, although on soils inclined to cap, a chain harrow may be preferable. Deep sowing must be avoided and the seed should be cut just on to the soil moisture and no deeper; $\frac{1}{2}$–$\frac{3}{4}$ in is ideal for most small seeds.

TABLE 25. Brassicas (Turnips, Swedes, Kale, etc.)

Approximate weight of seed (lb per acre) required for precision drilling

Spacing between seeds (*in*)	Seeds per foot of row	Width of row (*in*)						
		14	16	18	20	22	24	28
1	12·0	3·0	2·7	2·4	2·2	2·0	1·8	1·5
2	6·0	1·5	1·4	1·1	1·2	1·0	0·9	0·8
2$\frac{1}{2}$	4·8	1·2	1·1	1·0	0·9	0·8	0·7	0·6
3	4·0	1·0	0·9	0·8	0·8	0·7	0·6	0·5
3$\frac{1}{2}$	3·4	0·9	0·8	0·7	0·7	0·6	0·6	0·5
4	3·0	0·8	0·7	0·6	0·6	0·5	0·5	0·4

Mangold and Fodder Beet

Rubbed and graded seed $\frac{7\text{-}11}{64}$ in or $\frac{8\text{-}10}{64}$ in

Pelleted seed is approximately four to five times heavier than rubbed and graded seed.

Approximate weight of seed (lb per acre) required for precision drilling

Spacing between seeds (*in*)	Seed per foot of row	Width of row (*in*)						
		14	16	18	20	22	24	28
1	12·0	16·0	13·5	12·0	10·8	9·8	9·0	8·0
2	6·0	8·0	6·8	6·0	5·4	4·9	4·5	4·0
2$\frac{1}{2}$	4·8	6·4	5·4	4·8	4·4	4·0	3·6	3·2
3	4·0	5·4	4·5	4·0	3·6	3·3	3·0	2·7
3$\frac{1}{2}$	3·4	4·6	3·9	3·5	3·1	2·8	2·6	2·3
4	3·0	4·0	3·4	3·0	2·7	2·5	2·3	2·0

This Table is suitable only as a guide to the amount of seed required, as the weight of seed varies from year to year. There are also large differences in seed size between the various brassica crops which affects the weight of seed required per acre. These Tables slightly over-estimate the amount of seed required in order to allow an adequate margin for ordering.

Chapter 8

THE TECHNIQUE OF 'DIRECT DRILLING'

ONE OF the most promising methods of crop production introduced in recent years is the technique of 'direct drilling', which in this chapter is taken to mean the drilling of the seed directly into the soil without prior cultivation of the soil to prepare a seedbed. The surface vegetation, which is usually grass, is first destroyed by spraying with a solution of paraquat—the most suitable herbicide for the purpose introduced so far. The drill slots are usually closed either by rolling or harrowing.

Clean leys and cereal stubbles present the most suitable opportunities for growing direct drilled crops.

This technique should not be confused with the 'direct drilling' of certain horticultural crops such as cabbage and brussels sprouts, in which the seed is drilled straight into the field where they are to produce a crop, as opposed to the traditional method of raising the plants in a plant bed and then transplanting them.

In the United Kingdom 'direct drilling' is now commercial practice with the major cereals, wheat, barley, oats and rye, grown for grain or fodder, and with a number of brassica crops which can be sown thickly and do not require singling. They include kale, rape, and stubble turnips.

The direct drilling of turnips and swedes for bulbs is in a somewhat earlier stage of development, although the technique has proved to be highly satisfactory on substantial commercial acreages. With these crops the two major difficulties are to secure really even distribution of a small amount of seed and achieve a rapid and complete coverage of the ground by the tops. Grass and clover seeds have also been established commercially by direct drilling.

So far maize has only been direct drilled experimentally in the United Kingdom but in the United States of America it is direct drilled commercially on a very large scale. In that country, direct drilling of maize into a 'cover crop' of rye, which is first desiccated by spraying with paraquat, has proved to be preferable to conventional cultivations on slopes liable to severe soil erosion. The roots of the cover crop of rye, grown between two maize crops when the ground would otherwise be unoccupied, bind the soil and prevent erosion in heavy storms. The area suited to maize in the U.S.A. has been increased considerably by this technique.

ADVANTAGES

Direct drilling gives yields as good as those obtainable from crops grown by ploughing and conventional means and yet at the same time offers a number of important advantages. The virtual elimination of disturbance of the soil surface largely precludes the germination of annual weeds such as fat hen, knotgrass, and charlock (Plate 12) although a few annual weeds do occasionally grow from the drill slots or where the soil surface is otherwise disturbed.

Measures for the control of weeds after crop emergence are therefore usually unnecessary. This is a particularly important point, because the cost and effectiveness of herbicides in brassica crops grown by conventional means leaves much to be desired. The rapid drying out of the soil which occurs during ploughing and seedbed preparation immediately before sowing in hot summer weather is precluded. Soil moisture is conserved very well and there is much less likelihood of a patchy take or delayed germination.

Direct drilling is no cure for prolonged drought and should not be undertaken on dry soil unless rain is imminent.

The partial replacement of traditional cultivations by direct drilling could well exert a most beneficial effect on soil structure. Undisturbed soil compacts much less readily than soil which has previously been cultivated. The open 'structure' which has been built up by fibrous grass and cereal roots is left in position, while the virtual absence of cultivation tends to minimise losses of soil organic matter. Provision of efficient drainage is essential and periodic subsoiling or moling is still likely to be highly beneficial on some soils; direct drilling combined with periodic subsoiling when necessary appear to offer a promising combination.

Utilisation of direct drilled crops in situ is much more efficient. The firm soil surface greatly reduces the amount of poaching that occurs (Plate 12) and cows remain clean, making life much easier for the cowman. The incidence of sore teats and lameness, typical of dairy herds grazing badly poached fields of traditionally grown kale, is lessened.

Much less labour is required to grow direct drilled crops. Using a prototype 13 coulter triple disc drill to direct drill kale at Seale-Hayne, the whole operation, consisting of a single spraying, spreading fertiliser, direct drilling and rolling was completed in just under $1\frac{1}{2}$ man hours per acre (average size of field drilled 7·8 acres). In contrast, up to 9 man hours are usually required to grow kale for folding when ploughing and conventional cultivations are employed. On contract work the larger 15 coulter triple disc production model will cover 3–4 acres per hour according to size and shape of field.

The resultant saving of labour is only part of the story. In busy times such as haymaking and harvest, the ever-shrinking supply of labour on many farms does not permit extensive areas to be ploughed, cultivated and sown to catch crops such as kale and rape by conventional methods. In these circumstances, the choice is likely to be between direct drilling and not growing a crop at all rather than between two different methods of cultivation.

With a greater output per man hour, direct drilling allows much larger acreages to be sown at the optimum time, which is likely to result in improved yields and in greater acreages being planted. The time lost in the turn round from one crop to the next is also minimal.

DIRECT DRILLING BY CONTRACT

Direct drilling by contract removes most of the labour problem at planting and allows intensive cropping to be carried out when labour and capital are scarce. Most of the drills suitable for direct drilling that have been developed so far are relatively expensive single-purpose machines and except for syndicates or individual growers with large acreages, are at present best considered to be contractors' tools. Tying up capital in a machine used on a small acreage would be an unwise investment for most farmers.

Operators drilling large acreages are able to reach a higher degree of proficiency in the use of their drills than a person who only drills a small acreage each season. The successful development of reasonably priced multi-purpose drills may well alter the position considerably. Development work is in progress and already 'farm made' adaptations have started to appear.

Contract drilling appears generally to have worked well, although there is occasional evidence of operators trying to sow too large an area with a single drill. Farmers employing contractors should still supervise all operations personally. It is most unwise to leave all details in the hands of the contractor's driver; failures have occurred for this reason.

CHOICE AND PREPARATION OF THE SITE

To achieve success with direct drilling, it is necessary that the herbicide should control all unwanted vegetation rapidly and completely. As with all other methods of seedbed preparation, satisfactory soil conditions are required for drilling and for adequate root penetration. The selection and preparation of the site are therefore matters of fundamental importance. The technique provides no answer whatever to poor soil conditions or inefficient husbandry.

The most suitable soils are friable light to medium loams. Heavy soils, especially with autumn drilling, often prove to be difficult and tend to give the least satisfactory results. Poorly drained sites are useless. The soil must be free draining and not subject to surface waterlogging. Prior compaction of the soil by poaching with livestock or tractors restricts root development, giving a poor stunted crop.

In some instances where poaching of pasture prior to drilling has been severe, the root development of the subsequent kale crop has been observed to be confined to within $1\frac{1}{2}$ inches of the soil surface. In fact, one of the difficulties in following Italian ryegrass for early bite with direct drilled kale is the likelihood of stock poaching the ground in the early months of the year. If this should occur, the best course is to plough the ground and employ a spring fallow to restore the condition and tilth of the soil before drilling the next crop in the conventional way.

Very stony soils may only be planted with drills fitted with tines— disc or rotary drills are unsuitable. Success is also unlikely where infestations of perennial weeds such as docks, nettles and couch grass or old matted turf and excessive trash are present; they must be dealt with first.

If a site is initially unsatisfactory for direct drilling, it need not remain permanently so. Many sites, which in their current condition are unsuitable, could readily be improved to meet the required standards. The selection of the site, together with the necessary remedial measures, should be put in hand long before direct drilling is to take place—a year or more if possible. Any soil acidity should also be corrected for the previous crop. With direct drilled brassica crops it is essential that the pH of the top two inches of the soil should be between 6·0 and 7·5.

Careful preparation

Careful preparation of the site in the period immediately preceding spraying is of prime importance. The basic principles underlying the preparation of the site are the same for all crops, although the timing and details may differ somewhat. As paraquat acts only on the green parts of the plant, it is necessary that there is adequate green leaf present to receive the spray. At the same time, there should not be so much vegetation that the spray is unable to penetrate.

When spraying grassland, the ideal length of grass is between two and three inches. Whenever possible the best results are obtained by spraying the young regrowth after a cut of silage, when all first growth has been removed. Taking a late hay cut before direct drilling brassicas is undesirable, as large quantities of grass seed may be shed. If wet weather follows, this seed is able to establish and then competes severely with the young crop, a serious matter in the case of some of the slower establishing types such as dwarf thousand head kale and swedes.

The sward should also be open and free from trash. The presence of trash shields the grasses from the spray, so that patches remain unkilled. The seed may also become trapped in the trash rather than being deposited in the drill slots and thus fail to establish, a difficulty which has occurred on occasion with the 3D triple disc drill.

On the other hand, tined drills are inclined to block badly in trash. Trash or litter may also hold enough paraquat to be phytotoxic to young brassica seedlings. These may emerge very rapidly in warm moist weather and so come in contact with the paraquat before sufficient time has elapsed for it to be inactivated by the sunlight. Severe chlorosis of the leaves and stunting of the plants results (Plate 13). The trouble may also occur on trashy cereal stubbles. The old fashioned hay rake, where available, is an excellent tool for removing loose material, alternatively it may be burnt.

Obtaining a good kill on a grazed sward presents rather more difficulty and the results are sometimes less effective. Sheep grazings are often superior to those of cows, especially after two or three grazings, where the dung pats may shield the grass from the herbicide. If dung pats are present they should be spread and the sward mown over, leaving a stubble about two inches long which is allowed to produce the requisite amount of green regrowth, before spraying.

When following wet weather or with lax grazing, the presence of excessive soiled grass and trash may prevent the satisfactory penetration of the spray.

Cereal stubbles

Cereal stubbles infested with rhizomatous grassy weeds are quite unsuitable for direct drilling—they should be cleaned. Spraying with paraquat followed by straw burning is the best preparation for direct drilling. With a good burn, the result is a clean trash free surface and less risk of paraquat residues. If the straw is required, the shortest possible stubbles, free from spilled straw, should be left and allowed to green up if necessary. Any loose straw or trash remaining after baling or burning should be raked up and burned before spraying. The importance of a short well-prepared stubble which is free from trash or litter cannot be over-emphasised—it is vital.

Shed corn competes with the catch crop and may even suppress it in patches (Plate 13) so that yield is reduced. The problem can be minimised by harvesting as soon as the grain is ripe and by setting the combine correctly. Excessive delay in combining is the surest prescription for a carpet of shed corn, particularly in varieties which are inclined to neck or shed as soon as they are fully ripe.

When moist conditions favour the rapid germination of the shed corn, delaying the spraying by 7–10 days allows much of it to germinate before spraying. Delaying the application of paraquat will be of no advantage in a dry time and may lose the benefit of timeliness in drilling that is offered by the technique. Self-sown cereal plants may be controlled in brassicas only by spraying with $2\frac{1}{2}$–3 lb ai dalapon in 20–25 gall water when the crop has 2–4 true leaves. The best control for shed corn in Italian ryegrass is hard grazing and perhaps topping until the Italian ryegrass is well established. If not kept in check the barley can smother the young ryegrass seedlings.

Severely lodged cereal crops often present a particularly difficult problem. The amount of shed corn can easily be enough to stifle any crop trying to compete with it and the amount of trash may make direct drilling impossible unless a really good stubble burn can be obtained. Traditional cultivations followed by conventional drilling should be used in such a situation.

Spray Application

The success of the whole operation of direct drilling is dependent on efficient spraying; paraquat is a contact herbicide and really uniform application is essential.

The ease with which a sward may be desiccated depends on the type of sward and the time of year when spraying occurs. Old grassland, leys containing cocksfoot and fields sprayed before mid-June present the greatest difficulty, often showing very considerable recovery after a single spray application. In contrast, open swards of Italian ryegrass sprayed after mid-June usually present little difficulty and 0·5 to 1·0 lb ai paraquat per acre in a single application is then normally satisfactory. Stubbles usually only require 0·5 to 0·75 lb while 1·0 to 1·5 lb ai paraquat is necessary for older pastures.

When considerable regrowth is likely to occur, the best results are obtained by splitting the dressing of paraquat into two and allowing an interval of approximately ten days between the two applications. As there should always be a green target, the precise length of interval depends on the speed with which regrowth occurs. Split applications are much more effective and economical. Three-quarters of a pound ai paraquat is given at the first application, which desiccates the upper layers of the sward and opens it up to allow complete penetration by the second application. The latter, when a further 0·25 to 0·50 lb ai paraquat is given, desiccates the base of the sward and burns off any regrowth that may have developed. The split application of 0·75+0·50 lb ai paraquat is particularly effective. When spraying occurs before mid-June all but the most open swards should receive the two separate applications.

Even with spraying after mid-June two separate applications would seem to be safest with older leys and also those containing cocksfoot. If a single application is given before mid-June, when the grass is still growing actively, 1·5 lb ai paraquat per acre is recommended; the cost is then increased to around £5.80 per acre.

The precise timing of the spray application depends on the crop that is to follow. In the case of autumn and spring sown cereals more time is usually available for preparation of the site, as a quick turn round is by no means so critical. With an autumn sown cereal such as rye, grass, weeds and shed corn on stubbles should be allowed to green up before spraying.

On fields to be sown with spring corn, spraying occurs later but should be completed before Christmas to allow winter frosts to produce some surface frost mould; this is valuable for covering the seed. If creeping bent is present spraying must be brought forward to October.

Drilling

It is necessary to allow an adequate interval between spraying and drilling. If insufficient time is allowed, there is a danger that the crop seedlings may come in contact with traces of paraquat which are still herbicidally active on the trash or herbage. Normally these residues will be broken down by sunlight in a few hours but they can persist for several days when shaded in dense trash. Brassica crops are not very sensitive to paraquat residues and trouble normally only occurs when seedlings emerge rapidly in dense trash.

The manufacturers recommend that, with a young open ley, brassicas may be direct drilled within 1–3 days of spraying but with older and thicker swards it is safer to allow complete desiccation to take place; 4–7 days is then usually necessary. Grasses and cereals, on the other hand, are highly susceptible even to traces of paraquat which may remain active on trash or herbage. These are usually lethal, so that the minimum interval between spraying and drilling should be increased to seven or preferably ten days if practicable.

Except for spring cereals, drilling should, where possible, be completed within about 14 days of the final spraying.

Drilling should only be undertaken when the soil is reasonably friable as when conditions are suitable for normal drilling.

Light soils may be drilled a few hours after rain provided the drill runs freely but heavier soils require a longer interval to drain. The drill should not 'smear' the sides of the slots, as this impedes penetration of the soil by the roots. Drilling the seed too deeply, which is all too easy with drills designed specially to penetrate hard soil, must be studiously avoided. This fault is most likely on relatively soft stubbles and a number of failures have been experienced for this reason.

Brassica and Italian ryegrass seeds should not be placed deeper than one inch, preferably $\frac{1}{4}$ to $\frac{1}{2}$ inch deep. It is most important that a series of careful checks on the depth should be made as drilling proceeds and particularly at the beginning. Soil conditions can alter while drilling an individual field.

If the soil is very dry, it is often best to postpone drilling until either rain has fallen or there is a good chance of rain. Germination is most rapid when the seed is embedded in firm friable, moist soil. Rolling after drilling is usually necessary to close the drill slots and firm the soil around the seed, although compaction should be avoided. When the soil is drying rapidly, the roller should follow closely behind the drill; delay, even for an hour or two, may result in excessive drying out of the drill slots. Conversely, when drilling after rain it is better to allow a few hours for the surface of the disturbed soil to dry a little.

While conventional disc coulter type of corn drills have proved satisfactory on very light soils and relatively loose but trash-free stubbles, they do not usually penetrate uncultivated soil and specialised drills are therefore necessary. The major problem is cost.

Three types of mechanism

Direct drills employ three types of mechanism for soil penetration:
(a) Discs.
(b) Rotating blades.
(c) Tines.
The latter type has the advantage of lower capital cost and the ability to work under stony conditions.

Tined drills are normally quite useless in loose trash and a very high standard of site preparation is essential; stubble burning is an especially valuable precursor to this type of drill. Tined drills also tend to give greater soil disturbance, which is a marked advantage with autumn-sown cereals but with summer-sown brassicas may give trouble with annual weeds such as fat hen germinating around the drill slots.

Because the relatively low cost of tines and possibility of adapting existing tined implements for drilling, new developments are turning in this direction. Furthermore, the size and shape of the drill slot on any one type of drill may be varied readily to suit requirements simply by changing to a different type of tine. Replacement tines are cheaper than blades and disc units. The following drills are available commercially:

Plant Protection 3D drill: (Plate 14) This is a heavy triple disc drill specially constructed for direct drilling. The discs are independently sprung and they follow the minor contours of the land well. The drill slots are narrow, so that soil disturbance and consequent problems with annual weeds are minimal. Provided that a roller (Flat or Cambridge) follows the drill closely, soil moisture is conserved satisfactorily. Penetration of the soil is normally good but this drill is unsuited to extremely stony conditions. Fertiliser may be sown with the seed when required. A special attachment to the Plant Protection 3D and 3DC drills is available for sowing small seeds; rates down to 1 lb per acre may be achieved with brassica seeds.

Howard Rotaseeder: (Plate 14) This P.T.O. driven rotary drill cuts slots about an inch wide with rotating blades which are fixed to the rotors and does not follow small ground contours closely. Penetration and clearance of trash above the drill slots is excellent and there is little danger of seedlings coming into contact with material still contaminated with active paraquat.

Penetration of hard soil is very good indeed but blade wear may become excessive under very stony conditions, to which this drill is quite unsuited. The seed is deposited in a band of loose soil at the bottom of the slot. The slots cannot be closed readily and in the writer's experience, rain is usually necessary after sowing for germination to occur.

If the ground is too wet when drilling starts there is no soil to cover the seed and the sides of the slots can be badly smeared. More soil is disturbed than with the triple disc drill and occasionally trouble has been experienced with annual weeds germinating in consequence. This drill does not sow fertiliser with the seed.

Plant Protection 3DC drill: This drill is identical with the 3D model but the discs are replaced with tines; the two are interchangeable.

International Harvester Co. 6–2 Cultivator—drill: A heavy general-purpose machine with spring tines which performs well on hard soil

Close up view of single row maize harvesting attachment fitted to Claas "Jaguar" precision chop forage harvester. Note heavy waste left behind by unadapted "flail" type forage harvester which cut the headland.

The effect of type of forage harvester on the chop. Left of ruler: harvested by precision chop forage harvester; note fine chop and intimate mix of material. Right of ruler: harvested by unadapted "flail" type forage harvester, giving a totally unsatisfactory cut. Note the large pieces of ear and stem.

PLATE 12
Direct Drilling

Kale direct drilled in a ryegrass ley treat with paraquat. L sprayed 12th July a kale drilled on 14 July, 1971. Note u disturbed surface a absence of weed Photograph 21st Augu 1964, Pyrford, Surrey

Photos:
Imperial Chemical Industries Ltd.

The kale on the right was direct drilled into pasture after treatment with paraquat. The kale on the left was drilled into ploughed ground and is badly infested with charlock.

The Friesian herd of 71 cov belonging to Mr G. Howar Trevosper Pipers Poo Launceston, Cornwall, grazir marrow stem kale dire drilled into old pasture at 4l seed per acre with a 3D dr on 19th June, 1971. The fie was sprayed with paraquat c 15th June, 1971 and receive 132 units N, 66 units P_2O_5 ar 66 units K_2O per acre. Phot graph taken on 29th Noven ber, 1971 after 7 inches of ra had fallen during the mont Note the very small amou of poaching in the difficu conditions.

PLATE 13
Direct Drilling into cereal stubbles

...e effect of excessive loose
...sh and shed corn; note poor
...wth of rape (Essex Giant).
...ect drilled 27th August,
...1 into spring barley stubble.
...otograph 22nd September,
...1.

The effect of too short an interval
between spraying paraquat and
direct drilling where there is
trash or a thick sward. Field
sprayed 26th August, 1971, direct
drilled 27th August, 1971. Note
stunted chlorotic plants of Hungry
Gap kale which have come into
contact with trash carrying traces
of paraquat that was still active.

...eful crop of Emerald
... on the College
... at Seale-Hayne
...dy for feeding.
...ect drilled into
...ng barley stubble on
...d August, 1971 in
... rows; seedrate 5lb
... acre; manuring 80
...s N, 40 units P_2O_5,
...units K_2O. Sampled
... November, 1971.
...sh yield $10\frac{1}{2}$ tons
... acre; dry yield 22·6
... per acre.

The "3D" drill. Note discs which give minimal soil disturbance. This drill is also available with tines, and then designated the 3-DC.

PLATE 14
Specially designed drills

Photos: Imperial Chemical Industries Ltd.

The Howard Rotaseeder. Note wide slots and considerable soil disturbance.

Direct drilling kale with the Gibbs kale drill. Note the considerable soil disturbance, typical of all drills fitted with tines.

and in stony situations, giving greater soil disturbance than the two previous types; a useful point with autumn sown cereals for fodder. This drill is primarily for cereals; it will sow fertiliser with the seed.

Gibbs Kale drill: (Plate 14) This is a relatively light but specialised brassica drill with seeder units mounted on a toolbar frame. Soil penetration is obtained by 'pencil' type tines, which cause considerable soil disturbance. The seed is sown in wide rows.

Jones 519 drill: This is a dual-purpose machine, which is made up by mounting an 'Octopus' drill on a Konskilde 'Triple K' light cultivator. Kits are available to convert existing cultivators. The seeding mechanism is claimed to be suitable for cereals, grass and brassica seeds; fertiliser cannot be sown with the seed.

With the exception of the Gibbs Kale drill, all drills sow in rows 5–7 inches apart, but unwanted coulters can be blocked off as required. Where the seeding mechanisms are designed primarily for cereals an inert diluent such as minislug pellets, chick crumbs, seed size tapioca or kibbled maize is necessary when drilling small quantities of brassica seed. An unadapted triple disc drill, for instance, will not sow less than 4 and 8 lb of undiluted Brassica seed in 14- and 7-inch rows respectively; if lower quantities are attempted, 'mincing' of the seed results.

Diluents must be very thoroughly mixed with the seed and not too much of the mixture should be put into the box at one time, as seed and some diluents are inclined to separate with the shaking of the drill. Even then, the seed tends to fall somewhat unevenly, creating a number of gaps and bunches. However, the precise proportion of diluent/seed, which usually varies from 1/1 to 4/1 by weight, depends on the seed rate and on the setting of the drill, which should be set to sow as little as possible without mincing either seed or diluent. However, the most effective solution is likely to be the use of more sophisticated seed mechanisms.

Drill calibration

To obtain a stationary calibration check, the ends of the seed delivery tubes which are to sow seed are removed from the coulters and inserted in clean empty bags. The driving wheel is jacked up clear of the ground and with the drill in gear, the number of turns of the driving wheel that would be necessary for the drill to traverse $\frac{1}{10}$ acre are counted. A chalk mark on the tyre gives a point from which to count. The number of turns necessary for $\frac{1}{10}$ acre may be calculated as follows:

$$\frac{484 \times 9}{\text{sowing width of} \times \text{circumference of}}$$

sowing width of   ×   circumference of
drill (feet)   wheel (feet)

The weight of seed or seed plus diluent multiplied by 10 gives the rate per acre.

Calibration in the field is often more realistic than a stationary calibration. For this, seed or seed and diluent is placed in the hopper and a run of a few yards is allowed to ensure the drill is feeding before the seed collection bags are firmly attached to the ends of the seed tubes.

The drill is then towed over the measured distance necessary to cover $\frac{1}{10}$ acre, which is obtained by dividing 4,356 by the sowing width of the drill in feet. The weight of seed or seed and diluent is then multiplied by 10 to give the weight sown per acre.

On some farms the seed has been mixed with compound fertiliser and sown through the manure hopper. Although rates as low as $\frac{3}{4}$–1 cwt per acre of concentrated granular fertiliser have been used successfully, there is still a danger of phytoxicity and only inert diluents are currently recommended. If it is essential to use fertiliser as a diluent, up to 1 cwt granular superphosphate is least likely to cause trouble. Seed rates of direct drilled brassicas tend to be rather higher than where precision or semi-precision drills are used on conventionally prepared seedbeds.

The following rates (excluding diluent if required) are currently recommended:

	lb/acre	row width (in)
Kale (Marrow stem)	3–4	14
(Maris Kestrel)	2	14
Kale (Thousand Head)	$1\frac{1}{2}$	14
Swedes or turnips for bulbs ⎱ Thousand Head Kale + ⎰	⎰ 1 ⎱ $1\frac{1}{2}$	14 14
swede	$\frac{1}{2}$	14
Rape	4–5	7
Stubble turnips	2–4	7
Italian ryegrass (clipped seed)	18–20	7 drilled 2 ways
Forage rye	200	7

Slugs

Slugs can cause sporadic but serious trouble with direct drilled crops in the seedling stage. Incidence is erratic and does not follow any fixed pattern. When the crop is emerging, a useful routine measure is to lay down a number of 'slug traps' consisting of a small handful of slug pellets in the afternoon, and to count the dead slugs the next morning. If only one or two are present there is no problem but if the slugs are numerous, say 20 or more per trap, the field should be treated with either 28 lb per acre of proprietary slug pellets or a mixture of 28 lb bran, and 1 lb metaldehyde per acre.

The bran is first moistened and the metaldehyde powder is then thoroughly incorporated. This must be kept out of reach of children and domestic animals as it is poisonous. Small or minislug pellets may also be used as a diluent and mixed with the seed at 7–14 lb per acre, according to row width; 7 lb for 10–14-in rows, 14 lb for 5–7-in rows.

Metaldehyde is known to have weaknesses when incorporated with the soil. Control is improved to some degree by cultural methods. The absence of trash and closing the drill slots by rolling or harrowing is helpful.

Fertiliser Application

The absolute necessity for generous fertiliser treatment of direct drilled crops has been clearly demonstrated with commercial crops and in all field experiments undertaken to date. Shortage of nitrogen is a good deal more critical than with conventionally grown crops and at present the general assumption is that direct drilled crops require an additional 20 units of nitrogen. While normal dressings of phosphate and potash are generally considered to be adequate, deficiencies are likely to occur where leys are direct drilled into grass which has not been previously intensively manured and stocked.

As with lime, phosphate deficiencies should be corrected early in the winter before it is intended to sow the direct drilled crop, so that there is ample rainfall to wash the phosphate into the soil. Half a ton of high grade slag is a useful remedial measure to be followed by an application of a suitable compound fertiliser just before the crop is drilled. Slag is especially valuable for swedes.

Sometimes the slag encourages vigorous growth of white clover, which may compete unduly with the swedes. As the clover regenerates rapidly after treatment with paraquat, it is best controlled by spraying the pasture with mecoprop at least three months before drilling.

All phosphate and potash must be applied at or before drilling but the timing of the nitrogen application depends on the success of the grass kill. Provided that the surface vegetation has been properly killed, trials have shown that the best response is obtained where all the nitrogen is applied at sowing time. The only exception is, of course, an autumn-sown cereal where the bulk of the nitrogen application is invariably delayed until the spring.

Conversely, if recovery of the grass is likely to occur, which sometimes happens when brassicas are direct drilled into a ley which has only received a single dose of paraquat, the nitrogen application is best delayed until the brassica crop has 3–4 true leaves. If the nitrogen is applied earlier, it tends to encourage vigorous regrowth of the grass, which swamps the crop. By delaying the application, the brassica plants are large enough to make full use of the nitrogen and the position is then reversed.

Apart from the foregoing differences, the principles governing the manuring of direct drilled crops are the same as for those for the individual crops grown conventionally. General recommendations for direct drilled crops may be summarised as follows:

Crop	Base dressing Units/acre			Top dressing Units N/acre
	N	P_2O_5*	K_2O	
Kale (May/June sowing)	90	90	90–120	50–80
Rape	70	70	70	30–50
Turnips and swedes for bulbs	25–50	70–120	70–120	NIL–30
Stubble catch crops	75–100	40	40	NIL
Rye for early grazing	20	50	50	60–70 (in early spring)

* Phosphate deficient soils should receive 10 cwt basic slag at the beginning of the winter prior to drilling.

Choice of Crop

The main opportunities for direct drilling occur after grass and cereal stubble. When cereals are to be grown, there is usually adequate time for efficient preparation of the site, although no time should be lost with forage rye. However, with the brassicas choice of crop in relation to the dates of sowing and utilisation are important (Chapter 4).

Cereals, even winter barley, are mostly cleared too late for sowing swedes and kales such as marrow stem, the hybrid kales and thousand head. These usually follow a silage crop. On the other hand, rape, turnips, fodder radish and Italian ryegrass produce very well when direct drilled into a cereal stubble. Preceding a stubble catch crop by a winter cereal ensures earlier sowing.

Chapter 9

THE PRINCIPLES AND TECHNIQUES OF WEED CONTROL

THE CONTROL of weeds has proved to be one of the weakest links in the production of fodder crops. All too often, in the past, root and green forage crops have been planted only to become infested with weeds and produce half a crop. The use of hand labour is now largely out of the question; even when adequate labour can be obtained it can be used more profitably in other directions. Many weed problems may be reduced substantially or even eliminated by a sound understanding of the principles underlying weed control and by the adoption of techniques appropriate to the weed problem in hand. Accordingly no apology is offered for the considerable length of this chapter.

The period during which crops are most vulnerable to weed competition lasts from approximately crop emergence to the time that the leaves of the individual plants meet together to form a canopy over the ground; a good leaf canopy is capable of suppressing any further weed growth while it maintains an adequate density. Heavy crops are smother crops and give a 'built in' weed control.

The problem of weed control thus resolves itself into two distinct but equally essential parts:

(1) The provision of field conditions that encourage vigorous crop growth. The crop may then form the requisite leaf canopy in the shortest possible time and maintain it throughout the life of the crop.

(2) Control measures to suppress or eliminate weed growth during the vulnerable period which occurs prior to the completion of the leaf canopy.

MEASURES TO ENCOURAGE RAPID CROP GROWTH

Rapid production of a leaf canopy is largely dependent on efficient husbandry. All factors which contribute to the speedy establishment of the crop and which increase yield, such as free drainage, good soil structure, an adequate supply of soil moisture, ample fertiliser, the choice of suitable crops and varieties, high quality seed and freedom from pests and diseases produce crops which compete vigorously with the weeds and form and maintain a dense leaf canopy. Conversely any factors such as drought, seed of low germinating capacity or too low a plant population, which delay or prevent the formation of a good leaf cover, favour the weeds. The effectiveness of the canopy in suppressing weeds depends on its density and on its duration.

The various fodder crops differ greatly in their initial growth rate and in the density of the canopy which they produce. Some crops such as rape and fodder radish show an extremely rapid initial growth rate and are highly aggressive in their early stages. Provided adequate nitrogen is given, these crops are capable of outgrowing all but the fastest growing weeds and weed control usually presents no great problem.

Maize, which ultimately forms a good leaf canopy, is extraordinarily sensitive to weed completion during the initial stages of crop growth and stringent measures to control weed growth in the early life of the crop are essential.

The root crops are also vulnerable to weed competition in their early life but, unlike maize, they frequently fail to form an adequate leaf canopy later on. Weed control in root crops presents a considerable problem, especially when the tops are small or when a soil applied herbicide lacks persistency and weeds germinate before an effective cover is produced.

A well-grown crop of kale ultimately forms a dense leaf canopy which then suppresses further weed growth but the crop is sensitive to weed competition in its early stages. Kale, initially, makes quite vigorous lateral growth but can be outgrown in the vertical direction by erect weeds with a high initial growth rate such as charlock, runch and fat hen; heavy infestations then frequently smother the crop.

The effect of a liberal supply of nitrogen, whether applied as a dressing or artificial or already in the soil, is to favour the fastest growing or dominant species at the time the nitrogen becomes available to the plants. As the initial growth rate of most of the fodder crops is usually inferior to that of many of the weeds, a high level of soil fertility or liberal dressings of nitrogen in the early stages of crop growth usually tend to favour the weeds. In the absence of effective weed control measures, high fertility makes the situation worse and liberal applications of nitrogen should then be withheld until the weeds have been treated.

If nitrogen is applied before appropriate measures have been taken against the weeds, the crop may well be smothered even more quickly than it would have been if no fertiliser had been applied. Once the potential growth rate of the crop outstrips that of the weed, the position is reversed. Top dressings of nitrogen can then exert a considerable, if indirect, herbicidal effect; in particular the application of a late top dressing of nitrogen can be very beneficial in the case of mixed weed infestations in kale grown with traditional cultivations, as the herbicides suitable for use in kale after crop emergence will each control only a very limited range of weed species.

Vigorous crop growth greatly enhances the effect of herbicides. On occasions the herbicide may only check the growth of the weeds, so that the effect of the herbicide then depends on the vigour of the crop. If the crop is weak and offers little competition to the weeds, they are likely to recover and may still stifle the crop. Conversely, if the crop is growing vigorously, a check in growth is frequently fatal to the weeds and they are then suppressed by the crop. The more vigorous the crop and the denser the leaf canopy, the more complete is the suppression of the weeds likely to be.

MEASURES FOR WEED CONTROL

It is most unwise to omit weed control measures on the grounds that weeds are merely unsightly for a relatively short period in the life of the crop; although this may be true when only a few weed plants are present, there is ample experimental evidence to show that severe weed competition results in serious loss of yield. Even the yield of kale, which may ultimately outgrow the weeds, can be reduced by as much as 50 per cent by severe weed competition in the early stages of crop growth.

Type of weed

The essential feature of successful weed control is to match the control measures to the growth habit and pattern of development of the weed, so that the weed is dealt with in the most vulnerable part of its life cycle and subjected to treatments which it is least able to withstand.

It is all too easy to waste time and money on ineffectual efforts to control weeds. Attempts at weed control with heavy tractors and implements or with herbicides are quite ineffectual if the treatment is inappropriate to the weed or operations are undertaken when the weed is in a resistant phase. For practical purposes weeds may be divided into annuals and perennials; each group may be further subdivided into broad-leaved and grassy types.

Annuals complete their entire life cycle of germination, vegetative growth, seed production and death in a single year. Reproduction is by seed only, which is often produced in enormous quantities. Once shed, the seed of some species can remain viable in the soil for a very long time; the germination of a single seeding may be spread over a period of several years.

Weed species differ considerably in the time or period of the year during which the main germination of their seed occurs. This phenomenon, known as the periodicity of germination, is remarkably consistent for most species and occurs regardless of differences in the weather between one part of the year and another; seedlings of most weed species can be found at times of the year in which they do not normally germinate but their number is usually insignificant.

Certain weeds such as annual meadow grass, charlock, chickweed, corn poppy, groundsel, runch, scentless mayweed and shepherd's purse will germinate throughout most of the growing season, although there is a general tendency for germination to be more abundant in spring and autumn. The germination of most other important annual weeds is largely or totally restricted to a particular period of the year; in consequence, these weeds are frequently associated with particular crops or systems of cropping. This association occurs because the crops are sown before the main germination period of the 'problem' weed has been completed; the final cultivations for the crop cannot kill seeds which have yet to germinate.

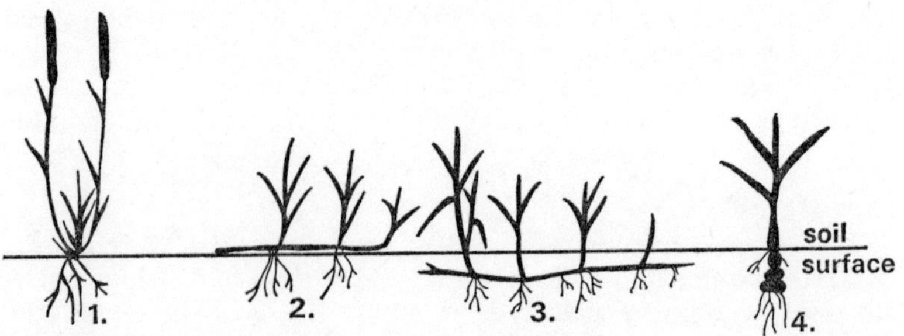

FIG. 6. Diagrammatic Representation of Growth Habits in Grassy Weeds.
Key
1. *Annual or perennial tufted grasses, eg, black grass.*
2. *Creeping or Stoloniferous perennial grasses, eg, creeping bent or water grass; rough-stalked meadow grass.*
3. *Rhizomatous perennial grasses, eg, couch.*
4. *Bulbous perennial grasses, eg, onion couch.*

TABLE 26. Period of Germination of Some Annual Weeds and Crops They are Likely to Infest

Main period of germination	Weeds	Crops in which weeds are most likely to prove troublesome	Remarks
Autumn	Blackgrass Winter wild oat Cleavers Corn buttercup Parsley piert	Winter cereals and winter beans.	Blackgrass normally germinates in the autumn but can be retarded by very wet soil: then germinates in spring when soil has dried out. Blackgrass and winter wild oat are much more troublesome where a run of winter cereals are grown.
Spring	Wild oat	Intensive spring cereal growing.	Can grow profusely in broad leaved crops but is not difficult to control therein. There is a small autumn germination.
	Black bindweed Knotgrass Redshank	Spring cereals, maincrop root and green forage crops.	These weeds do not germinate in late summer and autumn. Knotgrass usually finishes germinating by late May. Redshank starts later and continues well into July and may affect later sown crops.
	Fat hen Orache	Root and fodder crops sown before mid-June.	Fat Hen and Orache—readily killed by herbicides in cereals.
	Corn marigold Spurrey	Spring sown cereals.	Mainly weeds of light soils. Corn marigold is resistant to practically all herbicides except Dinoseb.
Summer	Black nightshade	Spring sown vegetables.	

If the same or similar crops are sown at the same time of year over a period of years, the weed is able to build up to infestation level when weed control measures are inadequate; some troublesome annual weeds and the crops with which they tend to become associated are shown in Table 26.

Perennial weeds differ from annuals in that they reproduce by seed and also by vegetative propagation; they do not die after seed production and their life cycle is thus extended over an indefinite number of years. Some species, such as docks, shed large quantities of seed from which plants establish readily; in contrast, the creeping thistle produces very little viable seed and propagates largely by means of underground horizontal stems or rhizomes. Once perennial weeds have become established it is necessary to kill their vegetative storage organs.

The growth habit of the various types of grassy perennial weeds, (Fig. 6), is of considerable importance, as it determines the effectiveness of the various measures for cultural and chemical control.

Preventive measures—general sanitation

Wherever practicable, weeds should be prevented from gaining access to the field. If any noxious weed does so, it should be dealt with immediately the first plants are seen and not allowed to build up to infestation level; a case in point is the wild oat. Clean seed is essential. Only certified or field approved seed (British Cereal Seeds Scheme) should be purchased; home grown seed requires cleaning before sowing. Corn fields should be rogued and the wild oat plants removed and burned at the stage when only a few plants can be seen. This practice is laborious but it is the most effective way of preventing a build-up to infestation level once wild oats have gained access to the field; if left, the wild oats soon become too numerous to remove by hand.

Particular attention should be paid to the ground under places where birds perch like electricity power lines. Purchased hay or straw and unheated farmyard manure are other potential sources of various weed seeds.

The prevention of seeding and the elimination of weeds in hedgerows and unavoidable waste places are important measures; weeds such as docks growing therein can seed profusely. Treatment with a herbicide such as mecoprop from a spray lance linked to a tractor mounted sprayer presents no difficulty and ensures grassy hedge bottoms and fence lines.

Grassy perennials such as couch and creeping bent may spread on to the headlands and be carried further into the field during cultivations; such spread may be confined by regular stubble cleaning of at least the headland, if not the whole field. Corners of fields are very wasteful of space and are ideal places for weeds to propagate; the removal of a short section of hedge adjoining the corner eliminates the problem and allows the rapid passage of wide implements and combine harvesters from one field to another. When the field is returned to grass the gap may be closed quickly by erecting a stake and wire fence.

Weeds may be controlled by cultivations, chemicals, rotational means or by any combination of these methods.

Rotational methods of weed control

During the 19th and early part of the 20th Century, weed control on arable land depended on the adoption of a sound rotation in which the 'fouling crops' (cereals) were alternated with 'cleaning crops' (the traditional root break); crop rotations were then legally enforceable on tenant farmers and were rigidly or even slavishly followed. The rigid enforcement of a pre-determined cropping plan was necessary, inter alia, to preserve the cleanliness of the land, as only relatively simple cultural methods of weed control were available and the timing of operations was critical.

The principle underlying the use of rotational means for weed control is that, by growing a range of different crops in succession, the environment and the timing of cultivations is changed from year to year. Conversely, when a single or very similar group of crops are grown over a period of some years, as in highly specialised cropping systems, the environment, management and timing of cultivations remain constant and particular types of weed infestation then tend to build up (Table 26).

The widespread plague of wild oats resulting from intensive cereal production is a case in point. This weed is highly sensitive to competition from broadleaved plants, both crop and weed. The disappearance of the root crop with its accompanying cultivations and the removal of any competition from fast-growing broadleaved weeds such as runch and charlock allowed the wild oat to multiply without hindrance. Competition was still further reduced by the growing of short strawed cereals.

Intensification of cereal growing meant that longer runs of cereals were taken and a build-up to infestation level could occur more rapidly. Earlier sowing, made possible by modern tackle, ensured that much of the cereal acreage was planted before the main germination period of the wild oat started; a rapidly increasing supply of seed, little of which was destroyed before the sowing of the cereal crop, freedom from severe competition from other plants and the capacity to shed much of its seed before or during harvest, gave the wild oat the perfect opportunity to build up to infestation level.

Other examples of weed infestations arising in a specific environment include blackgrass in winter cereals, docks in intensively managed leys and pastures and weeds such as annual nettles in intensive vegetable or market garden situations.

Chemical controls are now available for most weeds but the cost is frequently high and the ease with which specific weeds may be controlled varies from one crop to another. Established docks or thistles can be a real problem in root crops and direct drilled fodder crops but may be readily controlled by cheap herbicidal treatment in cereals and Italian ryegrass, whereas the wild oat may be controlled readily in a late sown root crop.

A sound rotation is an invaluable means of aiding weed control but the growing of any crop can only be justified if it is economically worthwhile. Thus, with the ever-increasing shortage of labour and rising production costs, root and green forage crops can only be justified because they fulfil an important function in their own right; there can be no question of growing them simply because they are useful in helping to control certain types of weed infestation.

Cultural control of annual weeds

Large numbers of weed seeds are usually present in most soils, even if they have been in grass for many years. While the soil surface is left intact these seeds remain dormant and present no problem but once the soil surface is disturbed, either by cultivations or by the poaching of livestock, conditions are produced which favour the germination of buried weed seeds, which then pose a considerable threat to the crop. Annual weeds, which reproduce entirely from seed, are therefore largely associated with arable cropping. Some old arable fields carry enormous populations of viable weed seeds.

Two basically distinct methods for controlling annual weeds may thus be adopted:

(1) The germination of weed seed may be prevented by eliminating cultivations, as in the technique of direct drilling (Chapter 8).

(2) Control is obtained by allowing or inducing weed seeds to germinate, as in conventional arable situations and then killing the seedlings or young plants by cultural or chemical means; the latter measures are ineffective against weed seeds which have not germinated before treatment.

Traditional methods of annual weed control rely entirely on cultivations; the two main traditional techniques relevant to root and green forage crops are the preparation of a 'false' or 'stale' seedbed and post-emergence inter-row cultivations. In the first case the weed seeds are induced to germinate by cultivations and successive crops of weed seedlings are killed by periodic harrowing before the crop is drilled. The success of the method depends on satisfactory weed germination, which in turn depends on adequate soil moisture and on carrying out the operation during a period of the year in which the weeds to be controlled will germinate. If these criteria are met, this method of weed control can be very effective and the crop then emerges on a relatively weed-free seedbed.

Unless it is specifically desired to bring weed seeds to the surface to germinate them, as in wild oat control, cultivations occurring in the latter part of the treatment should be as shallow as possible. To avoid the risk of soil compaction and loss of seedbed in periods of heavy rain the soil surface should only be left rolled after the crop has been sown. Weed control may frequently be improved and time may be saved by substituting the chemical version of the 'stale seedbed technique' which includes the use of a contact herbicide.

If properly carried out, post-emergence inter-row cultivations can provide a tolerably cheap and efficient method of controlling annual weeds growing between the rows in root and fodder crops grown with conventional cultivations; adequate dry weather is necessary. The secret of success is to catch the weed seedlings when they are only delicate threads and can be killed at a touch. The first pass of the tractor hoe should thus be made as soon as the rows are visible from the tractor seat; all the ground between the rows must be cut. Several passes of the tractor hoe are usually necessary before the crop plants are large enough to suppress any further weed growth. If singling is to be undertaken, a pass of the tractor hoe, immediately before singling, eases the work considerably.

If the first pass of the tractor hoe is delayed until the weeds are well established, a typical effect of prolonged wet weather, the hoe blades fail to cut many of the weeds and the technique is useless. Even if they are cut, some weeds such as chickweed re-establish only too readily in wet weather.

The real difficulty with crops grown in wide rows lies in controlling the weeds close to and within the rows. On most farms hand labour can only be used economically for removing unwanted plants from a precision drilled stand.

The alternative to inter-row cultivations and hand hoeing is to use an overall chemical treatment. The difficulty is that even if the herbicide chosen is effective against the weeds present, the cost of treating the whole of the crop area with herbicides can become quite prohibitive, especially if it is technically necessary to use two expensive complementary treatments. The technique of band spraying described in the section on weed control is a possible answer to the problem and is well tried in the sugar beet crop.

Cultural control of perennial weeds

There are no essential differences between the control measures required for annual and perennial weeds while they are both in the seedling and young plant stages; once perennial weeds become established, the basis of control has to be radically altered, as it then becomes necessary to destroy their vegetative storage organs.

In practice, the choice of treatment depends on whether the perennial weeds to be controlled belong to broadleaved or grassy species.

The control of broadleaved species such as docks and thistles presents no difficulty in efficiently sprayed cereal crops. A run of two or three successive cereal crops which are sprayed with a suitable herbicide will usually control most species of perennial weed.

The older weed control measures such as bare and bastard fallows are therefore no longer applicable for the control of broadleaved perennial weeds. Perennial broadleaved weeds in grassland may also be controlled by chemical treatment, although the task usually involves greater difficulty and more expensive chemical treatments.

Perennial grassy weeds are unaffected by the herbicides that can be used in growing cereal crops. With the exception of couch grass in maize, all cultural and chemical treatments must be undertaken when the ground is unoccupied by a crop. Any attempt at controlling perennial grassy weeds after a crop has been sown is likely to be totally ineffective.

Present-day land values and costs of production do not permit land to remain unoccupied by a crop for any great length of time. Except perhaps for really filthy heavy land on which it is desired to carry out spring drainage operations, the traditional bare fallow is obsolete and couch must, perforce, be controlled after a crop such as hay, cereal silage or cereals for grain has been taken. Growing winter barley allows an earlier start to be made with the cleaning operations and increases their efficiency.

Two quite distinct approaches may be employed for the eradication of couch by cultural means:

(1) Cultivations are employed to desiccate the plants. On light soils the rhizomes are brought to the surface, where they can be dried out in the sun. On heavy soils they are baked in the clod.

(2) The rhizomes are exhausted by a series of cultivations. The initial cultivations promote the sprouting of dormant buds. Subsequent cultivations kill the rhizomes by smashing off the regrowth every time new shoots are produced.

Desiccation is brought about on the lighter and freer working soils by breaking up a hay or cereal stubble to a depth of several inches, preferably with a heavy cultivator or chisel plough and then by bringing as many of the rhizomes as possible to the surface by harrowing or scuffling; the rhizomes are then allowed to desiccate in the sun. Periodic harrowing is continued to assist desiccation and bring any further rhizomes to the surface; harrowing continues until autumn rains stop further work.

Mouldboard ploughing should be avoided as the couch is best left on the surface throughout the winter; the succeeding crop is sown in the spring with 'minimal cultivations'. The couch must be brought to the surface to desiccate. The mere moving the soil with heavy implements may even worsen the situation. On heavy land the soil is burst up deeply with one or more passes of a very heavy cultivator and then worked to bake the clod as quickly as possible.

The technique of desiccation works extremely well in dry summers and in dry areas but is virtually useless under wet conditions. Chemical or chemical-assisted treatments are then greatly superior. Stubble burning and cheap chemical treatment are frequently a great help in clearing the land of trash before operations start.

The essential feature of the 'exhaustion' technique is to keep smashing the new regrowth before it has time to replace the nutrients that have been withdrawn from the rhizomes. If done properly, the rhizomes are then completely denuded of nutrients and the plant dies.

Rotary cultivations are particularly effective with this technique, although implements such as heavy disc harrows may also be used. The first pass with the rotary cultivator should be made in a low forward gear with high PTO revolutions to give a short 'bite' and cut the rhizomes into the smallest possible pieces. The work should be deep enough to undercut all the rhizomes. Subsequent passes should be 4–6 inches deep and are given as soon as the regrowth is 2 inches high or each new shoot has 2 leaves; an interval of 2–4 weeks is required between each pass.

Three passes are generally required for average situations; two passes may suffice on light land whereas four may be necessary on heavy land. One or more passes given on a single occasion, as in seedbed preparation, are worse than useless for couch control; the rhizomes are merely broken up, distributed more widely in the soil and propagation is accelerated. The 'exhaustion' technique may be applied at any time between late spring and the end of the cereal harvest, although treatment must start early enough to permit completion before the ground becomes too wet in the autumn. The exhaustion principle is also applied in the chemical treatment of couch with paraquat.

Chemical Weed Control

It is essential to differentiate between the terms 'herbicide' and 'herbicidal treatment'. The former refers only to the chemical which is used and whose effect is determined by the way and circumstances in which it is used. A given chemical may then give a totally different result when the circumstances are varied. The term 'herbicidal treatment' includes not only the chemical but the manner and circumstances in which that chemical is used and is much more informative. There are normally three basic components to any herbicidal treatment—the herbicide, the weed and the crop.

Herbicidal treatments may be either total or selective. Total or non-selective treatments are applied with the express object of killing all vegetation present and can only be safely applied to agricultural land when it is free from a crop as in the pre-sowing stage. The use of paraquat for turf destruction or the application of herbicides for the control of grassy perennial weeds between crops are suitable examples. Selective treatments are designed to suppress weeds without damaging the crop.

Herbicides may be applied either to the foliage of the weeds or to the soil. Foliar applied herbicides may be divided into contact and translocated types. Contact type herbicides affect only the parts of the plant with which they come into direct contact and are not moved away from the site of application to any great degree. They are therefore unable to affect inaccessible or protected storage organs and are normally of no value against perennial weeds unless repeated applications of the herbicide are made, as in the treatment of couch in cereal stubbles by paraquat.

A single application of a contact herbicide to perennials merely removes the foliage which soon regenerates. The use of contact type herbicides is generally confined to the treatment of annual weeds. Contact herbicides are only fully effective when the plants are really evenly covered; spraying technique is highly critical. A translocated herbicide is absorbed into the interior of the plant and is transported within the plant to parts remote from the point of application. Protected shoots and underground storage organs are thus penetrated and this type of herbicide is effective against perennial as well as annual weeds. Efficient spraying of translocated herbicides is still necessary but provided each weed plant receives some spray, the very uniform coverage essential with contact herbicides is not needed.

A 'residual' herbicide is applied to the soil. The herbicide may kill the roots by direct contact or it may be absorbed and translocated to other parts of the plant. In either case, a soil applied herbicide continues to affect the roots of germinating seedlings for some time after application; the duration of the effect varies according to the length of time the herbicide remains active in the soil. Soil applied herbicides are thus known as 'residual' herbicides.

Classifying herbicidal treatments

The most satisfactory method of classifying herbicidal treatments is according to the phase of crop development during which they are applied.

Pre-sowing treatments are applied before the crop is sown.

Pre-emergence treatments are applied after the crop has been sown but before crop emergence.

Post-emergence treatments are applied after the crop has emerged.

In each case, contact, translocated and residually acting herbicides may be used. Herbicides may be applied to the whole area to be treated, or overall spraying or their application may be restricted as in 'band' and 'directed' application.

With 'band' spraying the herbicide is applied to the soil surface or emerged crop and weeds as a band some seven inches wide running along each row and leaving a strip between each pair of rows to which no herbicide is applied. This strip is tractor hoed in the normal way. This compromise technique eliminates the need for hand work in weed control and at the same time reduces the cost of the chemical to about one-third of that of an overall application. The cost of the additional equipment can only be justified when there is a substantial acreage to be treated. Application may be made pre- or post-emergence of the crop.

'Directed' applications are made to kill weeds between the rows after crop emergence and is sometimes referred to as 'chemical hoeing'. Either foliar applied or residual herbicides may be used. Specialised equipment is necessary and the technique is not widely used for fodder crops.

Herbicidal treatments may now be classified as follows:

1. Pre-sowing or pre-planting	⎧	contact foliage
2.	⎨	or translocated foliage
3.	⎩	or residual
4. Pre-emergence	⎧	contact foliage
5. *	⎨	or translocated foliage
6.	⎩	or residual[1]
7. Post-emergence	⎧	contact foliage[1,2]
8.	⎨	or translocated foliage
9. *	⎩	or residual

Not applicable to fodder crops.

1. Band application may be used in some fodder crops.
2. Directed application may be used in some fodder crops.

M

The above classification is very useful but in some cases herbicides do not fall conveniently into a single category. Thus some 'residual' herbicides such as atrazine may also have 'contact foliar' properties while a 'contact foliar' herbicide such as paraquat may possess systemic properties and be translocated to a considerable degree under certain conditions such as darkness or low light intensity.

Pre-sowing treatments

Contact foliage: Herbicides used in this type of treatment have little or no residual effect and only persist in the soil or treated foliage for a very short time. Provided that the treatment of the weeds has been completed, the subsequent crop may be sown shortly after the final spray application. Paraquat is the best known herbicide used in this context.

Pre-sowing contact treatments may be used:

(i) To control annual weeds on a prepared seedbed immediately before sowing.

(ii) To clear land ploughed in the previous autumn which has become foul with annual weeds, plants from shed cereals or tufted grasses.

(iii) The destruction of a grass sward prior to ploughing or direct drilling.

(iv) The control of grassy weeds in cereal stubbles. This treatment may be given in one or more applications with or without cultivations, depending on the type of weed-grass to be controlled. Pre-sowing contact foliage treatments do not control broad-leaved perennial weeds such as docks and thistles.

Translocated foliage: Herbicides such as dalapon or amino-triazole are used to control couch grass and some broadleaved perennial weeds when the ground is unoccupied by a crop. These herbicides have some residual effect and an appropriate interval must be left after treatment to avoid damage to the succeeding crop. The choice of treatment and its timing depend on the weed to be controlled, on the soil conditions and on the cropping programme. Dalapon, for instance, is unsuitable for foliar application to couch on shallow soils and winter cereals cannot be planted after an autumn treatment.

Residual: In pre-sowing and pre-planting treatments relatively persistent herbicides which are mainly root-absorbed, are incorporated into the soil at some time before the crop is planted. The timing of the treatment depends on the persistence of the chemical, the dose required and the tolerance of the crop to the chemical.

When it is necessary to control perennial weeds, such as an infestation of couch grass, a relatively heavy dose of a persistent herbicide such as TCA is incorporated into the soil. The dose generally persists for several months and a relatively long period must elapse before it is safe to sow a susceptible crop. The requisite interval may be shortened by growing a crop with a greater degree of tolerance or by reducing the dose of the chemical, which course may involve some sacrifice in the degree of efficiency of weed control.

Residual herbicides may also be mixed with the surface layers of the soil shortly or immediately before sowing for the control of germinating annual weeds. The crop must then be tolerant of the herbicide. If not, sowing must be delayed to allow time for the chemical to be dissipated in the soil. Thorough and even incorporation of the chemical is necessary. Volatile chemicals should be incorporated immediately after application. Di-allate (wild oat control) and Trifluralin (control of broadleaved annuals) are used in this way. It is necessary for soil applied herbicides to be held in the soil long enough for the crop to produce a dense foliage cover. If the herbicide is lost from the soil too quickly, weed control is likely to be ineffective.

A limited number of residual herbicides may also be applied shortly or immediately before sowing or planting tolerant crops for the control perennial weeds like the use of atrazine to control couch grass in maize. Atrazine is normally used as a pre-emergence application at a lower rate to control annual weeds. At the higher rate necessary for couch control it is much more persistent and its use in this way necessitates the growing of two successive crops of maize.

Pre-emergence treatments

Contact foliage: This type of treatment is not persistent and only kills weed seedlings that have actually emerged at the time of application. The treatment can be highly effective if all the weeds emerge in a single 'flush' but may be virtually useless if any substantial number of weeds germinate after application. The treatment does not rule out the need for cultivations after crop emergence.

Pre-emergence contact sprays were first applied to slow germinating crops such as sugar beet and onions and the success of the treatment was highly dependent on correct timing. The crop was drilled and spraying was then delayed until 2–3 days prior to the time of estimated crop emergence, so that the maximum number of weeds had emerged before treatment took place. In dry weather many of the weed seeds failed to germinate whereas the advent of prolonged wet weather usually precluded spraying at the right time and crop emergence occurred before spraying could take place.

The 'stale seedbed technique' was developed for use in quick germinating crops, notably swedes and kale. The seedbed is worked to a fine tilth and then left for 10–14 days to allow the weed seeds to germinate; the seed is then drilled with minimal soil disturbance. Spraying can occur a day or so either side of drilling. Spraying after drilling allows time for more weed seedlings to emerge and tends to kill more weed seedlings but if the weather is unsettled it is much safer to spray shortly before drilling. The stale seedbed technique depends on an adequate germination of weeds before spraying. Once the seedbed is prepared for weed germination wet weather sometimes causes the soil to run together and the seedbed is lost. The last operations before spraying or drilling should therefore be harrowing rather than rolling; a chain harrow is ideal for the purpose.

Translocated foliage: This treatment is rarely used as most translocated herbicides leave some toxic residues in the soil which may damage the crop seedlings.

Residual: Pre-emergence residual treatments involve the application between sowing and crop emergence of a residual herbicide which kills germinating weed seedlings over a relatively long period. It is usually best to apply the herbicide as soon as possible after drilling. Suitable herbicides are generally of low solubility in water and persist in the surface layers of the soil. Some large seeded crops, such as beans, may be drilled deeply so that their roots do not come into contact with the herbicide but in most cases it is necessary that the crop is tolerant or resistant to the herbicide.

Pre-emergence residual herbicides are extremely dependent on moist soil and good growing conditions for success. A fine smooth clod-free seedbed is essential; cloddy seedbeds give patchy and uncertain results. Dry soil conditions at the time of spraying and afterwards render the herbicide ineffective; this risk is greater with crops such as mangolds and maize which are sown in late spring or early summer. The activity of atrazine applied to maize can be improved in dry conditions by shallow incorporation in the soil.

The availability and consequent uptake of soil applied herbicides by weed seedlings depends on the soil type. Clay and humus particles tend to adsorb and inactivate residual herbicides, so that the doseage rate generally has to be increased as the clay content of the soil rises and on organic or peaty soils. The degree of adsorption varies from one herbicide to another; some residual herbicides, like atrazine, are very severely affected and are unsuitable for residual application on soils high in organic matter.

A number of 'residual' herbicides also possess contact properties; in addition to their longer lasting residual affect they also kill many types of weed seedlings and sometimes, as in the case of atrazine, even kill young plants of some species. This additional property is of considerable value for the post-emergence spraying in crops of maize grown on organic soils or where spraying has been unavoidably delayed until after the emergence of crop and weeds. When applied at the normal rate for annual weed control pre-emergence residual herbicides do not control established perennial weeds; these should be controlled before sowing the crop.

Post-emergence treatments: Post-emergence foliar applications possess several major advantages over pre-emergence residual treatments. Foliar applications are independent of the state of the tilth and soil type and are unaffected by a dry soil surface at spraying or by dry weather afterwards; in fact dry weather for some hours after spraying is a pre-requisite for success with most foliar applied herbicides. The majority of the weeds have generally emerged before post-emergence foliar treatments are given.

Post-emergence foliar applications provide the normal treatment for all broad-leaved weeds in cereals and by far the greatest part of the huge acreage sprayed with herbicides is treated in this way. A very wide range of post-emergence foliar herbicides is now available for use in cereals. In contrast, very few foliar applied herbicides are available for the post-emergence treatment of broadleaved weeds in the root and brassica green forage crops and most of these are limited in the number of weeds which they will control.

Contact foliage: The selectivity of post-emergence contact foliar applications depends partially on the amount of herbicide retained by the crop and by the weed respectively. The spray solution tends to run off leaves with a waxy coat such like those of kale but is retained by weeds with a hairy leaf surface such as runch and charlock; conversely corn marigold possesses a smooth leaf with a thick waxy coat and is extremely difficult to kill by selective foliar applications.

The retention of the herbicide may be improved greatly by the addition of a 'wetting' agent, which lowers the surface tension of the spray solution, causing the droplets to spread out and stick to the leaf surface; the disadvantage is that the selectivity of the herbicide is reduced and the crop may then be severely damaged.

The success of post-emergence foliar contact herbicides is generally dependent on volume rate, droplet size, weather conditions and the stage of growth of the weeds. The tendency is to use medium to high volume (very low volume = <5 gal/acre; low volume = 5–20 gal/ acre; medium volume = 20–60 gal/acre; high volume = 60–100 gal/acre) and small spray drops to achieve the requisite even coverage. Warm, dry weather during and immediately after spraying gives by far the best results.

Annual weeds tend to be more susceptible to post-emergence foliar contact herbicides in their early stages of growth and the range of weeds that a given herbicide will control can frequently be widened by spraying at the earliest stage that crop growth and weed emergence permits.

Translocated foliage: In contrast to contact applications, the selectivity of overall translocated treatments depends on the difference between crop and weed in what happens after the herbicide is absorbed into the plant. Initially, the outwardly visible effects are slow to develop but become more obvious as the herbicide increasingly interferes with the growth of susceptible plants. Translocated herbicides are much less dependent on volume and drop size than contact herbicides.

Translocated foliar herbicides are totally unsuitable for use in root and brassica green forage crops and the selective control of broad-leaved perennial weeds in these crops is impossible. Perennial broad-leaved weeds must either be dealt with in the pre-sowing period of the fodder crops or they must be controlled selectively in previous grass or cereal crops; the latter course is generally the most economical. Maize is sensitive to damage from translocated herbicides and residual treatment is usually to be preferred.

Residual: Treatments are currently inapplicable to fodder crops.

Choice of Method of Weed Control

The ultimate choice of method of weed control is by no means a simple matter, as many factors have to be considered. Weed control may be greatly improved in the longer term by the adoption of a varied cropping sequence and by sound forward planning; particular weeds may then be dealt with efficiently and more cheaply in the appropriate situation, rather than left until they have become an acute embarrassment.

Whatever the method of weed control, it must be suited to the agronomy of the crop and to the farming situation. Although factors such as soil, climate, current weather conditions and the crop in which the weeds are to be treated are very important, the choice is frequently restricted by the ease with which the treatment may be carried out, the cost of the treatment and the value of the crop.

The available resources vary greatly from one farm to another. Arable farmers growing a wide range of crops usually possess ample equipment and are experts in carrying out sophisticated herbicidal treatments; their staff are also highly experienced. Many of the crops on this type of farm have a high cash value. If it is not already on the farm the purchase of additional equipment such as band sprayers can usually be justified. In contrast, the stock farmer frequently possesses little arable equipment and is unfamiliar with the more sophisticated herbicidal techniques. It is also often difficult to justify expensive chemical treatments for fodder crops.

Cultural techniques for weed control depend to a large degree on fine weather conditions and are more relevant to the drier parts of the country. Their performance can frequently be improved by the incorporation of a cheap chemical treatment which widens their application considerably. Prolonged cultural treatment can result in the ground remaining uncropped for an excessive period. Even a relatively costly chemical treatment may then prove to be a worthwhile investment provided that its use does in fact result in higher crop output.

Small farm staffs and high wages are making purely cultural methods of weed control progressively more unattractive. The application of chemical treatment, including direct drilling on contract, can reduce the labour requirement to a minimum but the use of these techniques must be looked upon as buying one's labour in another form. Their use must therefore allow existing labour to be diverted to other productive work, permit the labour bill to be reduced or result in a higher farm output. Alternatively the use of a given herbicidal treatment may be the only way open to a farmer with little labour of growing a particular crop.

The choice of treatment is also influenced by the intensity of the weed problem. Sound rotational and cultural techniques such as pre-sowing harrowing and delayed drilling are most useful preventive measures and may well be more economical when weed incidence is relatively light. When really heavy weed infestations occur over a large area, there is usually little option but to invest in chemical treatment. The increase in yield resulting from the removal of weed competition, the reduction in labour and the timely sowing of a larger acreage of the more profitable crops can fully justify the investment.

Spray application

The success or failure of a herbicidal treatment depends to a very large extent on the efficiency of the spray application. Good maintenance of the sprayer is essential. The machine should be thoroughly overhauled before being put away for the winter and worn or defective parts must be replaced. Nozzles should be renewed every season on farms where any substantial spraying programme is undertaken. The cost of new nozzles is relatively small compared to the value of the chemical that passes through them. Worn nozzles result in inefficient spraying and waste of chemical.

The sprayer should be calibrated before use according to the maker's instructions and the output checked periodically. The correct pressure, volume rate and type of nozzle should be used for each job. Excessive pressures must be avoided as they increase the drift hazard. The positioning of the nozzles and height of the spray bar should be adjusted so that the spray fans or cones meet a few inches above the top of the crop or target to be sprayed.

The manufacturer's instructions on mixing should always be followed. Clean tap water should be used; dirty water block filters and can reduce the effectiveness of the chemical, especially that of paraquat. A large covered tank with twin ball cocks at the filling point speeds turn round considerably when the sprayer has a high capacity filling pump; the tank is filling while the operator is away spraying.

A proper filling routine should be established. Gloves and face shields should be used when handling dangerous types of concentrate and protective clothing must be worn when the chemical requires. Empty drums should be washed out and the rinsings tipped into the sprayer during filling and then safely disposed of or buried to avoid contaminating rivers and water supplies.

Thorough washing out of the sprayer is essential when changing chemicals, as some crops can be severely damaged by even the smallest traces of chemicals remaining in the machine. The tank should be emptied and the sprayer cleansed with ample clean water and spray cleaner. Washings should be sprayed out on waste land well away from ponds and water courses.

When spraying it is important to avoid overlap or missed strips; good marking is essential. The purchase of foam markers may be considered for larger acreages. Double spraying, where the second application is given at right angles to the first, is useful in situations such as stubbles where the wheel marks are difficult to see. Half the chemical is used at each pass. The second pass should be given immediately after the first.

Field conditions and stage of growth of crop and weed must be right. Spraying at the wrong stage can damage the crop severely or give an ineffective weed control. Herbicides are most effective when the weeds are making vigorous growth. Spraying in drought or cold weather gives slower results and is frequently less effective. Spraying wet foliage or spraying shortly before heavy rain is inadvisable with most foliar applied chemicals.

Drift is a major hazard and can do irreparable damage to susceptible crops. Broadleaved fodder crops and most horticultural crops, glasshouse crops, orchards and gardens are extremely susceptible and are readily damaged by even small traces of drift or vapour from the 'growth regulator' translocated herbicides such as

MCPA, 2,4-D and mecoprop. Spraying should only be undertaken in calm weather and never 'up wind' of susceptible crops; the possession of an adequate sprayer and equipment which gives a quick turn-round allows farmers to be more selective in their choice of weather for spraying. Excessive pressures and unnecessarily small spray drops increase the dangers of drift. In spite of every precaution accidents can happen and it is only prudent to possess adequate insurance cover against any third party claims likely to arise from damage by spray drift; the cover should be substantial, as claims for damage to glasshouse crops can involve very large sums of money.

WEED CONTROL IN ROOT AND GREEN FORAGE CROPS

1. Selective Control of Broadleaved Perennial Weeds

The most economic and satisfactory control of infestations is frequently obtained by ploughing, preparing a seedbed by conventional means and sowing a crop or run of crops that is sprayed with a post-emergence translocated herbicide such as a pure stand of Italian ryegrass and/or a cereal crop. A run of two or more sprayed cereal crops is highly effective against most species. Some species of perennial weeds, like docks, are more expensive and difficult to control in pasture. Pastures should always be sprayed during establishment to eliminate seeding docks.

DOCKS

(1) Control by ploughing: Bad infestations controlled by Italian ryegrass ley sprayed as early as possible during establishment, at least twice during summer and followed by a sprayed cereal crop (2, 4-D amine, MCPA, Mecoprop or 'broad spectrum' herbicides).

(2) Control in pasture in the year prior to ploughing or direct drilling.

(i) Asulam: Two applications spring and autumn.
Approximate cost £4.00 per acre.

(ii) Dicamba + mecoprop + MCPA mixture: One application April–mid-October. Second application may be necessary. Approximate cost £3.50 per acre.

(iii) Mecoprop: Cut docks to ensure good regrowth in July/August and spray when growing actively. Bad infestations may require two sprayings, say in May and again in July. Last spraying at least 3 months before sowing broadleaved fodder crop.
Approximate cost up to £1.40 per acre.

THISTLES

(1) Control by ploughing: As for docks.

(2) Control in pasture in year prior to ploughing or direct drilling: One to two applications of MCPA or MCPB/MCPA mixture.

NETTLES

(1) Control by ploughing: Nettles killed readily by efficient cultivations and sprayed cereal crop.

(2) Control in pasture in year prior to direct drilling: Nettles can be very troublesome with direct drilling and usually occur in clumps. Spot treatment with 2,4,5T or Mecoprop in spring and again in autumn; especially valuable in September or October before early frosts.

OTHER PERENNIAL WEEDS

Buttercups: May be controlled by spraying in cereal crops or in pasture. Bulbous buttercup is difficult to control in grassland and may give trouble with direct drilling. It shoots in the autumn and dies back after flowering in the spring. Autumn spraying may be preferable.

Horsetails, rushes and silverweed: Weeds of this type frequently indicate poor drainage. An investigation should be made before ploughing or direct drilling is undertaken.

2. Pre-Sowing Measures

GRASSY PERENNIALS

(1) Desiccation of pasture before ploughing: With 0·50–0·75 lb paraquat per acre 7–10 days before ploughing kills the sward and prevents creeping grasses, such as rough stalked meadow grass and creeping bent from invading the subsequent crop, especially winter cereals. The resultant stubble is relatively clean and is more suitable for direct drilling a fodder crop.

By allowing later ploughing the treatment also permits additional keep to be taken off the field. Use at least 0·75 lb paraquat ai for creeping bent. The grass should have 2–3 in young *green* regrowth at spraying as paraquat *only* acts on green material; not over 4 in. Spray in September.

(2) Stubble burning is an excellent way of clearing stubble and trash prior to stubble cultivations for the control of couch, ploughing or direct drilling fodder crops. A clean stubble prevents the blockage of tined implements or drills by trash and allows direct drilled seedlings to emerge on a trash-free surface without damage from paraquat residues.

For burning, apply 0·50 lb paraquat ai per acre across the field after combining. With regular swaths the nozzles over the straw may be shut off so that only the intervening ground is sprayed. A firebreak of cultivated land should be prepared around the headland. Firing takes place when desiccation is complete, usually 4–7 days after spraying and should be done into the wind on a dry day after the dew has cleared; mid afternoon is best.

If the straw is required, leave as short a stubble as possible, apply 0·50 lb paraquat ai and start cultivations as soon as desiccation has taken place. (For direct drilling see Chapter 8.)

RHIZOMATOUS GRASSES OR 'COUCH'

Serious infestations may be controlled by the following chemicals combined with appropriate cultivations, usually in a cereal stubble. Costs of treatment are usually rather high and can then only be justified with heavy infestations.

(1) Residual: TCA may be used on heavy infestations in single applications (30 lb/acre in autumn, 15 lb/acre in spring) or in split applications (2 × 20 lb/acre in autumn or 2 × 15 lb/acre in spring; the latter are less convenient but more reliable in some circumstances). All but the lighter soils are best treated in autumn. Cultivate stubble 5 in deep, spray and incorporate immediately. Rotavator-sprayers make an excellent job and no further incorporation is necessary.

If heavy rain falls after application a further cultivation should be given. Ploughing may occur six weeks after the single application. In wet autumns it may be impossible to carry out the final cultivations after the split application.

TCA may persist in the soil for a while. Post-harvest or autumn treatments may be followed by normal cropping (except winter cereals) but with a spring treatment (spring cereals are unsuitable) the following intervals should be allowed:

30 lb/acre: mangolds and fodder beet 8 weeks; kale, turnips and rape, 4 weeks.

15 lb/acre: mangolds and fodder beet 10 days; kale, turnips and rape, 7 days.

Approximate cost £3.20 to £8.40/acre.

(2) Translocated foliage: Dalapon (11·5–14 lb dalapon sodium/acre) or Amino triazole 8 lb/acre is applied to the foliage of couch grass which is growing actively. The stubble should be rotavated or cultivated as soon as possible after harvest to break up the rhizomes and induce active growth. Sufficient time should be allowed for a good cover of young leaf (4–6 in long) to develop. In the North there is frequently too little time after harvest for the treatment to be effective. Deep ploughing (minimum 8 in) not less than two weeks after spraying is essential to prevent regeneration. These treatments are unsuited to very shallow soils.

Approximate cost of dalapon £4.70 per acre.

(3) Contact foliage: 'Low cost' stubble control of couch may be obtained by a series of at least three applications of *paraquat* (contact herbicide) at low dosage rates (0·12 lb paraquat ai acre). The treatment prevents the couch from building up food reserves and exhausts the rhizomes. The stubble is burned or cleared of straw and trash and is rotavated or cultivated as soon as possible after harvest to break up the rhizomes and induce the buds to grow. Rotavation is preferable. The couch is sprayed every time the regrowth reaches the two-leaf stage. A wetting agent must be added to the tank and not less than 25 gal of water should be applied. This method has given useful results in trials. Some farmers replace a spraying with a good scuffling in dry weather. Approximate cost of three applications £1.40 per acre.

STOLONIFEROUS GRASSES

(1) Creeping Bent or **Watergrass:** Heavy infestations of watergrass can reduce the yield of cereals considerably. Ploughing in a heavy infestation without chemical treatment is extremely difficult. Even if successfully achieved, reinfestation usually occurs and may result in so much trash in the stubble of the succeeding cereal crop that the success of a direct drilled stubble catch crop may be jeopardised. The stubble may be grazed if required and is sprayed with paraquat at 0·75 lb ai per acre in October before the leaves on the stolons or runners lose their green colour and become harder to kill; the grass should not be longer than four inches at spraying or the spray will not penetrate. Recovery may result if spraying occurs too early.

Approximate cost £2.80 per acre.

(2) Rough stalked meadow grass and other annual and perennial tufted grasses: The stubble may be sprayed with paraquat at 0·50 lb ai at any time as required, although it is better to wait until October if no cultivations are undertaken previously. The stubble should be well greened before spraying. These techniques, Stoloniferous Grasses 1 and 2 are especially valuable in wet harvests and under west country conditions. Ensuing cultivations are much easier.

Approximate cost £1.90 per acre.

Paraquat: Applied at 0·25 lb ai per acre may be used to 'freeze' the growth of grassy weeds until stubble cultivations can conveniently be undertaken. This technique is applicable in a wet spell or where harvesting operations necessitate a late start to cleaning operations. Allow the stubble to green up before spraying.

Approximate cost £0.95 per acre.

3. General Measures for the Control of Annual Weeds

PRE-SOWING CONTACT

Stale seedbed techniques, mangolds, kale, swede and turnips: The seedbed is fully worked and then left for 10–14 days when it is to be hoped that the majority of the weed seedlings will have just emerged. Spraying with a contact herbicide occurs 24 hours before or after drilling. Spraying first is safer if wet weather is likely to follow. Only weeds which have emerged are controlled.

Dosage rate depends on species and size of weeds to be controlled. These herbicides may also be used with mangolds and fodder beet.

Approximate costs per acre: dimexan £2.75; paraquat £1.90–£2.80.

PRE-SOWING RESIDUAL

Control of germinating wild oats; also black grass (early drilled crops only)—Di-allate: Kale, cabbage, swedes and turnips; must be incorporated into the top 4 in of the soil immediately after spraying by thorough cultivations. May be used in conjunction with other soil applied herbicides.

Approximate cost £3.00–£3.75.

TCA applied to mangolds or kale at least five days before drilling at 7 lb ai to control wild oats. The herbicide must be thoroughly incorporated into a level seedbed as deeply as possible.

Approximate cost £1.50.

Selective Weed Control in Individual Crops

MAIZE

Chemicals have generally superseded inter-row cultivations for weed control in maize; the latter are ineffective within the row and are likely to damage the prop roots.

Residual treatment—(1) Annuals: Atrazine applied at 1–1½ lb ai per acre as soon as possible after the light harrowing following sowing. Under dry conditions a further light harrowing after spraying is likely to accelerate the effect of the herbicide. Atrazine may also be applied as a post-emergence treatment but seedling weeds should not be more than ½–1½ in high according to species. No crop other than maize should be sown before late autumn; winter oats are sensitive to atrazine and should not be sown in the autumn after treatment.

(2) Control of couch: Either (a) *Atrazine* applied at 4 lb ai per acre 2–4 weeks before seedbed preparation or (b) as a split application with 2 lb atrazine ai applied when couch starts to grow actively in the early spring and ploughed in within one month, followed by further application of 2 lb atrazine ai within 7 days of drilling the maize. No crop other than maize may be planted for 18 months after application; this treatment therefore involves growing a second crop of maize. Winter oats should not be planted after the second crop.

(3) When pre-emergence treatment has been omitted or is unsuitable as on soils with a high organic matter content, mixtures of 8–12 oz atrazine + 8 oz 2,4-Damine may be applied as soon as weeds emerge after sowing provided the maize is at least 3 in high.

OTHER CEREALS FOR FODDER AND ITALIAN RYEGRASS

A very wide range of herbicides is available for the control of practically all broad leaved weeds in these crops. Selection depends on the weeds to be controlled and can be made with the aid of the Weed Control Handbook Vol. 2 and from manufacturers' recommendations.

MANGOLDS AND FODDER BEET (Table 27)

Pre-emergence residuals: A fine, even seedbed and adequate moisture is essential for the success of soil acting herbicides.

Post-emergence contact: Phenmedipham should be applied to mangolds after crop and weeds have emerged but before the weeds have passed the recommended stages. Knotgrass must be sprayed while still in the cotyledon stage.

Nitrate of Soda: At $2\frac{1}{2}$–3 cwt acre in 100 gal water/acre for control of infestations of some annual weeds; mangolds with at least two true leaves. Hot, sunny weather is essential; this herbicide will not control fathen. Emergency use only.

CABBAGE AND KALE (Table 27)

Pre-planting residual: Trifluralin is applied while the seedbed is being prepared at any time during a 14-day period immediately before drilling or planting; it must be incorporated into the top $1\frac{1}{2}$ in of the soil within 30 minutes of application either by rotary cultivator or cross-harrowing. There should be an interval of at least 4 months after application before cereals or grasses are sown as a subsequent crop. Suitable on sandy loam, loams, silts and clay loams.

Pre-emergence residual: Propachlor is applied to transplanted cabbage when the weeds are just starting to emerge and to drilled cabbage and kale shortly after drilling.

Post-emergence contact: Desmetryne and SMA should be applied as soon as the growth of the crop permits (*Desmetryne*—kale and drilled cabbage at least 3 true leaves and 5 in high; transplanted

TABLE 27. Some Herbicides that May be Used to Control Broad Leaved Annual Weeds in Root and Brassica Green Forage Crops Grown with Conventional Cultivations

	Mangel and Fodder Beet	Kale	Flat Poll Cabbages (transplanted)	Swedes and Turnips	Approximate Cost (overall spraying) £/acre
Pre-planting/sowing					
Trifluralin	X	O	O	O	£5.50
Pre-emergence residual					
Lenacil	O	X	X	X	£4.20–£9.50(bd)
Propachlor	X	O	O	O	£5.60
			(seedbeds)		
Propham + Fenuron + Chlorpropham	O	X	X	X	£7.70(d)
Pyrazon (e)	O	X	X	X	£3.75–£11.25 (bd)
Post-emergence contact					
Desmetryne	X	O	O	X	£2.20–£3.40 (c)
Phenmedipham	O	X	X	X	£7.70
Propachlor	X	X	O	X	£5.60
			(after transplanting)		
Sodium monochloracetate (SMA)	X	O	O	X	£2.20–£2.50 (c)
Sulphuric Acid (BOV) (f)	X	O	X	X	—
Nitrate of Soda (f)	O	X	X	X	—

KEY: O = suitable; X = must not be used.

(a) Increased rate of application and price for greater weed population.
(b) Rule of application varies according to soil type: increased rate and cost on heavier and organic soils if suitable.
(c) Rate of application and cost varies according to dose.
(d) Cost may be cut to approximately one-third by band spraying if equipment is available.
(e) Pyrazon may also be applied as a pre-sowing or post-emergence treatment in some circumstances; it may also be applied in mixture with di-allate in a pre-sowing treatment for wild oat control.
(f) Emergency measures only. Obtain contractors quotation if available.

TABLE 28. The Susceptibility of Some Annual Weeds to Herbicides Listed in Table 27

| | Applied to Soil | | | | | Applied to Foliage | | | | | |
| | | | | | | Seedling Stage | | | Young Plant Stage | | |
	Lenacil	Pro-pachlor	Propham +Chlor-propham +Fenuron	Pyrazon	Tri-fluralin	Des-metryne	Phen-medipham	SMA	Des-metryne	Phen-medipham	SMA
Annual meadow grass	S	S	S	S	S	R	R	R	R	R	R
Annual nettle	F	S	S	S	F	R	S	S	R	S	F
Annual sowthistle	S	F	R	—	R	S	R	F	—	R	R
Black bindweed	S	R	S	S	S	F	S	S	R	R	S
Black nightshade	R	F	R	S	S	F	R	S	F	R	R
Charlock	S	R	S	S	R	F	S	S	R	S	F
Chickweed	S	F	S	S	S	S	S	F	F	F	R
Cleavers	R	S	S	R	R	F	R	F	R	R	F
Corn marigold	S	S	R	—	R	R	—	R	R	—	R
Fat hen	S	F	S	S	S	S	S	R	S	S	R
Fumitory	S	R	S	F	F	S	S	R	F	F	R
Groundsel	F	S	S	F	R	F	S	F	R	R	F
Hempnettle	—	S	F	S	S	S	S	S	F	S	F
Knotgrass	S	F	S	S	S	R	F	R	R	R	R
Mayweed	S	S	S	S	R	S	R	F	—	R	R
Orache	S	R	S	S	S	S	S	R	F	F	R
Pennycress	S	R	S	S	R	F	F	S	R	R	F
Poppy	S	—	S	S	F	S	S	R	F	F	R
Redshank	S	F	S	S	S	S	S	S	F	R	S
Runch	S	R	F	S	R	F	S	F	R	S	R
Shepherd's Purse	S	S	F	S	R	R	S	S	R	R	F
Speedwell	R	S	F	F	S	F	S	S	—	F	—
Spurrey	S	S	S	S	F	S	S	S	S	F	S
Wild Oat	R	R	F	R	S	R	R	R	R	R	R

KEY: S = Susceptible; F = Fair or variable control; R = Resistant.

cabbage at least 2 weeks after planting. SMA—kale 2–5 true leaves; drilled cabbage 3–4 true leaves; transplanted cabbage as soon as established); weed growth becomes much less susceptible as the weeds grow older.

Sulphuric acid: (Applied to Marrow Stem or Thousand Head kale as 8 per cent v/v solution at 100 gal/acre to control annual weeds when the kale is between the 2–6 leaf stage. A 10 per cent v/v solution or undiluted sulphuric acid (BOV) may be applied at the 4–6 leaf stage at 20 gal/acre. The kale must be growing vigorously at the time of treatment. A dressing of nitrogen helps the crop to recover from the scorching produced by the treatment. An acid-resistant sprayer is required.

This treatment is usually a contractor's job. Always add the acid gradually to the water, stirring continuously. NEVER add the water to the acid. Emergency use only.

SWEDES AND TURNIPS (Table 27)

Pre-sowing residual: Trifluralin, as for cabbage and kale.
Pre-emergence residual: Propachlor, as for cabbage and kale.
Agricultural Chemicals Approval Scheme
Users are advised to purchase only chemicals included in the list of 'approved products for farmers and growers' published annually by the Ministry of Agriculture. Products that are not included in this list may give satisfactory results in some cases but are unreliable in others.

Manufacturers' instructions must always be read carefully before use and be properly observed.

Chapter 10

THE ECONOMIC ASPECT

IT IS a relatively simple matter to compare the costs and financial returns of two arable crops grown for direct sale. The costs of production can be assessed with reasonable accuracy and the yields and precise cash returns are known as soon as the produce is sold. The apportionment of some 'overhead' or 'fixed' costs and the valuation of by-products may present some difficulties but they need not distort the picture unduly.

The choice of crops for direct sale is thus a fairly straightforward financial decision; such is not the case with fodder production. The growing of fodder crops or the conservation of hay and silage are only the first stage in the complicated process of milk and meat production; even when successfully produced the fodder still has to be converted into a marketable product and sold at a profit.

It is extraordinarily difficult to measure the 'yield' of a fodder crop. Fresh or green yield bears no relationship to the dry matter or nutrient content, as shown in Table 29. Although the value of the information may be improved greatly by expressing yield in terms of dry matter or calculated starch equivalent and cost as cost per ton of dry matter or cost per ton of starch equivalent, these are still unsatisfactory measurements of the costs and suitability of the foods under comparison. They take no account of the quantity of the amount of each food the stock will either readily or may safely consume nor do they give any indication of the differences in animal performance likely to result from the consumption of the various foods in differing proportions.

The value of any comparisons on a nutrient basis is further reduced if the yield or quality of the fodder varies from that assumed in the calculation. The yield and cost per hundredweight of starch equivalent of hay in Table 29 is based on a yield per equivalent acre of good quality hay (ie the total recorded output of hay and grazing is converted into terms of good quality hay).

TABLE 29. Forage Crop Costs and Yields Per Acre

Reproduced from 'Costs and Efficiency in Milk Production 1968–69'.
Crown Copyright, reproduced by permission of The Controller, HMSO.
The starch equivalent values assumed are those shown in Rations for Livestock Bulletin No. 48, HMSO, London.

Crop	No. of costs	Average cost £ per acre	Average Yield per acre		Average cost	
			Yield Tons	Starch Equivalent cwt	£ per ton	£ per cwt of starch equivalent
Hay	341	18·23	3·1[1]	22·2	5·88	0·82
Grass silage	108	21·82	10·8[1]	26·0	2·02	0·84
Cabbage	7	72·38	24·3	46·1[2]	2·98	1·57[2]
Kale—grazed	83	17·84	15·1	28·8	1·18	0·62
Kale—carted	17	36·89	12·5	23·8	2·95	1·55
Rape	16	12·84	10·3	14·3	1·25	0·90
Mangels	20	62·10	31·1	38·6	2·00	1·61
Swedes and turnips	17	49·79	18·2	25·9	2·73	1·92

[1] Yield per equivalent acre.
[2] The SE of cabbage can either be the value taken 9·5 (open-leaved cabbage) or 6·6 (drumhead cabbage). If the latter value is used the yield of SE is reduced to 32·1 cwt per acre while the cost is increased to £2.25 per cwt of SE.

If the yield is left unaltered but the quality of the hay is only poor (SE 30) or even very poor (SE 25) the yields of starch equivalent fall to 18·6 and 15·5 cwt per acre, while the costs per cwt of starch equivalent rise to £0.98 and £1.18 respectively. A typical two-ton per acre cut of hay of average quality (SE 30) will yield about 12 cwt of starch equivalent per acre. The same limitations apply to the yields of nutrients from silage and the other fodder crops but the feeding values of roots and green forage crops tend to vary much less.

Quality is of prime importance when hay or silage form a large part of the ration. Low quality products may so reduce the intake of digestible nutrients by stock that productivity is reduced to almost nothing. Lambs and bullocks will not fatten and ewes and cows will not milk satisfactorily on a heavy diet of poor quality hay or silage; far from being cheap, such low quality foods are dear at any price for productive stock. A root or green forage crop may cost more, but if it produces out of season fodder and results in higher production or in a greater sale price for the produce, then the extra cost may be regarded as a worthwhile investment. 'Fill belly' hay and silage is best kept for dry stock or sold.

When a crop is grazed the total yield has little practical significance. The really important quantity, the proportion of the crop that is actually eaten by the stock (the utilised yield) is almost impossible to ascertain and one is forced back on to further assumptions.

The comparison becomes much more meaningful when yields and costs are given in terms of the output of the animals that actually consume the crop. Even then, the real criterion is the profit produced by the whole livestock enterprise as opposed to the yield or cost per gallon of milk or per hundredweight of meat. While quantity, quality and cost of the fodder exert a great influence on the profitability of the enterprise, their precise effects are extraordinarily difficult to define in exact terms. The size and nature of this problem goes far beyond the scope of a single chapter.

The limited purpose of this chapter is therefore to suggest ways in which the costs of fodder crop production may be kept to a reasonable minimum and to indicate the factors likely to determine the suitability of the various fodder crops to an individual farming situation.

COSTS OF PRODUCTION

Costs of production may be divided into 'variable' and 'fixed' costs. The 'variable' costs are those which are only incurred when the crop in question is grown; if the crop is foregone and the ground remains uncropped, these costs do not arise. The variable costs incurred in crop production include seeds, fertilisers, herbicides and insecticides, sundry stores (twine, canes and nylon thread for bird scaring, slug pellets, etc.), contract services and casual labour. These costs may be ascertained readily and the total set against the growing of the crop.

'Fixed' costs are the costs which are incurred whether a crop is grown or not and are carried regardless of the type of crop. 'Fixed' costs include rent, regular labour, depreciation and repairs of machinery and any sundry items which cannot be allocated specifically to any particular enterprise on the farm such as accountant's fees.

Paradoxically, there is nothing so fixed as the 'variable' cost and nothing so variable as the 'fixed' cost. If full yields and hence minimal cost per unit of production are to be obtained, there is little room for manoeuvre in the use of seeds and fertiliser other than by avoiding waste or uneconomic application. As yield is dependent to a very large extent on there being an adequate number of plants and a high level of fertility, it is false economy to reduce seed or fertiliser application to a level below the optimum for the situation in question.

Any further reductions in seed costs are generally trifling when compared to the potential value of a full crop and the use of very low seed rates or the purchase of 'cheap' seed are just not worth the risks they entail.

Fertilisers have become much more expensive since the removal of the fertiliser subsidy and there is a great temptation to reduce the level of application to save money. It is still just as essential as before to apply enough fertiliser to grow a full crop but much can be done to ensure that fertiliser is used to the maximum economic effect.

Nitrogen should still be applied generously in situations where it will give a good response—to crops grown after a long rung of arable crops, to early sown crops or after there has been heavy leaching of soil nitrogen in a wet winter.

The application should be reduced where the response is likely to be small—crops grown after rich turf, crops receiving heavy applications of farm yard manure or slurry or late sown crops. Soil reserves of phosphate and potash should not be allowed to become depleted or yields are likely to suffer. The wider use of the cheaper forms of insoluble phosphate such as basic slag and ground mineral phosphate in the wetter areas of the country could help to reduce costs.

We have become so accustomed to subsidised artificial fertilisers in recent years that there has been a tendency to regard the question of farmyard manure and slurry as an inconvenient matter of sewage disposal. Large amounts of plant food can be obtained from these sources. If livestock residues are properly conserved and distributed liberally they can frequently save considerable expenditure on artificials.

Casual labour, machinery, contractors' charges and herbicides are possible substitutes for one's own labour and machinery and are best considered in the context of the 'fixed' costs of labour and machinery.

The variable costs are easy to compile and may be set out readily 'on the back of an envelope'.

Specimen variable costs

Specimen variable costs are shown for a number of crops in Tables 30 and 31. The cost appropriate to the situation of the reader may be arrived at by substituting the cost of the actual seed, fertilisers, herbicides, stores, casual labour and contract services employed. When most of the work is done on contract, as with direct drilling, when there is a low labour content to growing the crop or when the crop is grown as a catch crop, the variable costs comprise the largest part of the total cost and form a useful basis for cost comparison in fodder crops.

When production costs have a high regular labour or farmer owned machinery content, variable costs become a less satisfactory basis on which to make a comparison.

Fixed costs

The allocation of fixed costs to the various crops requires the keeping of accurate time sheets and detailed costings; the preparation of such costings is best left to the trained economist. The level of 'fixed' costs varies so much from one farm to another that it is most unwise to transplant 'average' costs from one situation to another. 'Average' costs are also frequently a very unsound basis for inter-crop comparisons.

It is in the field of the so-called 'fixed' costs that the greatest economies can be made, especially in the expenditure on labour and on the costs of machinery depreciation and repairs. Rents, on the other hand, cannot be reduced and will probably continue to rise. The most effective approach to 'fixed' costs or 'overheads' is to prune them as hard as possible and to attain a really high level of output so that the remaining total is spread as thinly as possible over a really large volume of sales.

The major items in the 'fixed' costs category, which include rent or mortgage interest, labour costs and machinery depreciation and repairs, may also be looked at in the light of land, labour and

TABLE 30. Examples of Variable Costs Per Acre of Some Brassica Fodder Crops.

(Seedbed preparation is by conventional means unless otherwise stated (costs as at 21.6.72)

Stubble brassica crops, direct drilled	£	*Kale, direct drilled into ley or Italian ryegrass*	£	*Swedes, direct drilled into long ley*	£	*Swedes, drilled to a stand with herbicide*	£
Seeds							
5 lb rape	0.60	2 lb Maris Kestrel	2.00	1 lb ungraded seed	0.25	½ lb graded seed	0.30
or 5 lb rape kale	2.00	*or* 4 lb Marrow stem	1.60				
or 3 lb turnip	0.90	*or* 1 lb canson					
		+ ¾ lb swedes	0.70				
Fertilisers[1]							
3 cwt compound (25:9:9)	6.00	4 cwt compound (9:25:25)	9.00	5 cwt compound (9:23:18)	10.60	5 cwt compound (9:23:18) (with boron if required)	10.60
		3 cwt nitram 34·5% N.	1.58	1 cwt nitram 34·5% N.	1.60		
				10 cwt basic slag 14% insol. P_2O_5	5.30		
Herbicides							
Paraquat 0·50 lb (1 application)	1.94	Paraquat 1·00 lb (1 application)	3.88	Paraquat 1·25 lb (2 applications)	4.85	Paraquat 0·50 lb *or* Trifluralin	1.94 / 5.50
Sundry stores							
10 lb mini slug pellets to bulk seed to 14 lb	0.22	5 lb mini slug pellets to bulk seed to 7 lb	0.11	6 lb mini slug pellets to bulk seed to 7 lb	0.13		
Contract services[2]							
Direct drilling	3.50	Direct drilling	4.00	Direct drilling	4.00		
				Spreading slag	0.50		
Total variable costs							
12.26–13.66		19.27–20.57		21.43–27.23		12.84–16.40	

[1] *Fertilisers.* Prices based on June 1972 early delivery rebate rates. Basic slag is appropriate to the above crops on soils low in phosphate, e.g. Wales and is not always required for direct drilled swedes. Boronated fertiliser only used where necessary.

[2] *Contract services.* Contractors charges vary greatly. Direct drilling depends on the acreage to be covered and tend to be less for cereal stubbles. Some contractors quote below the stated rates. If spraying or precision drilling or mechanical harvesting are done on contract these charges should be included in the variable costs.

TABLE 31. Examples of Variable Cost Per Acre of Some Selected Fodder Crops

Seedbed preparation is by conventional means (costs as at 21.6.72). It is assumed no contract services are used.

Italian ryegrass (*Autumn sown*) *manuring includes early bite only*	£	*Forage rye for grazing*	£	*Whole crop barley for silage*	£	*Silage Maize*	£0.20/lb.
Seeds							
20 lb Ab. S.22	2.20	1¼ cwt Lovaszpatonai	5.25	1¼ cwt barley	3.13	40 lb @ 1400/lb	8.00
Fertilisers (June 1972 EDR prices)							
3 cwt compound (20:14:14)	6.23	2 cwt compound (17:17:17)	4.28	3 cwt compound (20:14:14)	6.23	6 cwt compound (20:14:14)	12.45
2 cwt Nitram 34·5% N.	3.16	1½ cwt Nitram 34·5% N.	2.37				
Herbicide				Broad spectrum	1.25	1½ lb atrazine ai	2.25
Sundry stores						Canes and nylon, say	1.50
Total variable costs	11.59		11.90		10.61		24.20
Remarks Italian ryegrass is frequently taken on for a cut of silage and followed by kale; a further 3 cwt 20:10:10 type of fertiliser would be given after the early bite.		Forage rye is normally only used for early bite and is followed by a second crop.		Cereal silage is normally followed by Italian ryegrass (undersown) or kale.		20:10:10 compound adequate where soil reserves of phosphate and potash are good.	

capital. These, together with a farmers' 'management skills' and his basic expertise' form a major part of his resources. Profitability depends to a large extent on the successful deployment and exploitation of these assets in the set of circumstances peculiar to the individual farm. The limiting factors, lack of management expertise, inadequate or poor quality land, shortage of labour or insufficient capital vary from one farm to another.

Use of land cropping systems

The value of most agricultural land is now so high that only by intensive farming can an adequate return be obtained on the capital locked up in this asset. If the farmer is a tenant rather than an owner-occupier, the value of the land is likely to be reflected in increased rental charges. High stocking rates combined with high utilised yields and maximum output are an economic necessity; these are affected greatly by the cropping system.

Three distinct methods of fodder crop production may be identified:

(1) The traditional 'root break' or 'maincrop' which occupies the ground for the whole or the greater part of the growing season.

(2) The 'catch crop', which is snatched between two maincrops when the ground would otherwise be unoccupied; the maincrops are usually grown for direct sale.

(3) The situation where two fodder crops are grown on one piece of land in the same year. Technically, one of the crops is the maincrop and one is the catch crop but which is which is frequently a matter for conjecture. Examples include cereal silage followed by kale, rape or Italian ryegrass and Italian ryegrass or ley followed by swedes, turnips, kale or rape.

Maincrops

Growing a fodder crop as a maincrop involves the total sacrifice of any direct income and the entire cost of growing and utilising the crop must then be met by the stock consuming it. Like the grass break, the fodder crop has to compete economically with other crops grown for direct sale.

Alternatively, when the fodder crop becomes a substitute for grassland, either for the whole or for a part of the season, its produce must then be equated to the grass that would have been produced during the actual period that the fodder crop occupied the land. Thus, if kale is sown in June or July, a cut of silage or hay may be taken first and the production of the kale is then comparable to the production of the grass aftermath. A comparison with the first and heaviest cut of hay or silage is frequently irrelevant. When a whole year's grass production is foregone to grow a crop such as mangolds or maize, these crops are in direct competition with the entire production of the grassland for the year.

Swedes grown in the north of England and Scotland, mangolds, cabbage and maize are the most important examples of maincrops. The economic viability of these crops depends to a large extent on the production of a really heavy or even massive yield of digestible nutrients and then utilising the crop with the utmost efficiency. Effective utilisation of this type of crop usually involves harvesting, storing and rationing at least part of the produce of the crop; utilisation of such crops in situ is frequently too wasteful unless part of the crop is first carted off the field.

'The effect of yield on cost per ton is clearly illustrated in the case case of swedes by the following example.* Assuming a variable cost of swedes at £20 per acre, yields of 20, 30 and 40 tons per acre will show a variable cost of £1.00, £0.67 and £0.50 per ton respectively. The yield of swedes and turnips shown in Table 29 (18·2 tons per acre) is about half that which should be obtained from a good maincrop and is more appropriate to a catch crop. Similarly, the average yields of mangold and cabbage are too low; either the yields of these crops should be increased greatly (50–100 per cent) or their production should be discontinued; liberal manuring and a very high standard of husbandry are essential with this type of crop.

Catch crops

The concept of catch cropping is quite different to that of main crop production; a yield of cash or fodder has already been obtained and the alternative would be for the land to lie idle until the next maincrop is planted. When the catch crop is well chosen and the standard of cultivation is high, no maincrop or part of a maincrop need be sacrificed. Most of the 'fixed' costs have already been met and therefore only the additional costs (the variable costs), incurred as a direct result of growing the crop can reasonably be charged against the catch crop. The crops normally grown, kale, rape, stubble turnips and Italian ryegrass have very low labour demands, especially when they are direct drilled.

Catch cropping is an excellent way of increasing the stocking rate and intensifying the farming system without sacrificing direct cash income or incurring undue labour charges. The concept is essentially a 'cheap' method of intensification with low labour input, low yields by maincrop standards and is usually combined with utilisation in situ. 'Cheap' or not, the crop must not be skimped; adequate seed and herbicide must be used and the crop must receive enough

---

*Extracted from Reid, J. and Hunt, R. W. T. 'Turnips a Declining Crop?' Rep. No. 134. Econ. Dept. W. Scot. Agric. Coll. pp. 31.

fertiliser to ensure the production of a worthwhile yield. All too often catch crops have been planted without any fertiliser at all; the result of such misguided parsimony is invariably half a crop or even total crop failure.

The system of catch cropping is not without its pitfalls. If the catch crop occupies the land too long or the soil is dried out, the yield or even the success of the following crop is likely to be jeopardised. It is essential that an advance decision is made by the management as to the relative importance of the maincrop and the catch crop; when income from the maincrop is the prime consideration, the maincrop must be given preferential treatment and the catch crop is subordinated in all respects.

The catch crop must be removed early enough to allow ample time for adequate seedbed preparation and timely sowing or planting of the maincrop. When a crop is likely to carry or perpetuate any pest or disease of an important maincrop, it is generally inadvisable to grow it as a catch crop in the same rotation.

On land where brassica crops such as swedes, brussels sprouts, cauliflower or cabbage are of prime importance, catch cropping with brassicas must be avoided. Crops such as rye or Italian ryegrass, which are not subject to clubroot, are a far wiser choice. Similarly most brassica catch crops can perpetuate beet eelworm and should be avoided on land where sugar beet is grown.

A further difficulty with catch cropping is that the clearing of the preceding main crop may leave too short a growing period for the catch crop to produce a worthwhile yield; such is the case with stubble rape or turnips grown after late harvested cereals or spring cereals in the north of England. The most economical course is then to abandon the idea of catch cropping or to plant a crop which is suited to the later sowing, such as rye or Italian ryegrass.

The growing of catch crops must also be equated to the number of stock on the farm, the level of production of other forms of fodder and the practicability of utilisation. There must be an adequate stocking density to justify the catch crop and it should result in a higher stocking rate then would otherwise have been carried. The stock must utilise it profitably and it should result in increased livestock or crop production.

If existing grassland is not fully productive, it may be much cheaper to produce the additional fodder by applying more fertiliser to existing grassland and by improving the standard of grassland management. Alternatively, catch crops can be grown as a substitute for grassland, thus allowing some of the less productive grass fields to be ploughed and sown with arable crops for direct sale.

Catch cropping has wide application, especially in the south of England where there is a relatively long growing season. Substantial areas of crops such as early potatoes, peas, beans and winter cereals are harvested while there is still a useful proportion of the growing season left; in these cases the productivity of the land could be raised considerably by following with an appropriate catch crop, provided, of course, that the land is free from couch grass. Catch crops can also prove to be most valuable in cases of emergency, as with crop failures or after a difficult hay harvest.

Double cropping

This system involves the growing of two fodder crops during the year so that, like the traditional root break or maincrop system, it involves the sacrifice of any cash income from direct crop sales. As with maincrops, the success of the system depends on a high rate of production of digestible fodder and its efficient utilisation.

In contrast to the usual maincrops such as mangolds, swedes and maize, which produce a heavy final yield of highly digestible material, the 'useful' yields of crops such as cereal silage and kale are limited and fall below a satisfactory level obtainable from a well grown maincrop. The limitation may be imposed either by the relatively early maturity of the crop, which is the case with whole crop cereal silage or by the need to limit yield to avoid the laying down of valueless, woody material in the stem, as with kale. Waste is also excessive when very heavy crops are utilised in situ. The philosophy is very similar to that underlying the utilisation of grassland, where several cuts or grazings of reasonably good quality are taken rather than one large cut of very indifferent quality.

The two crops in a 'double crop' system are capable of jointly producing a very heavy total yield of digestible fodder over the whole year.

Problems of labour, mechanisation and capital

When employing capital it is essential to invest it to the greatest advantage and thus obtain the highest 'net benefit'. The labour required for crop production may be provided in a number of different forms. The traditional method of using 'regular' or 'casual' hand labour for tasks such as singling and harvesting roots is disappearing fast.

The replacement of labour by capital, which is done through the purchase of farm machinery is expensive and, if outright purchase by a single individual is contemplated, the acreage covered by the machine must be large enough to justify the initial capital outlay, the depreciation, the 'cost of the capital' and the running costs; if not, other arrangements such as employing a contractor or forming a syndicate to share the machine should be entered into. A purchase is much easier to justify when machinery can be used for more than one crop and the cost can be spread over a wider base.

It is only too easy to lock up capital in buildings and machinery that could be employed much more profitably by investing it in productive assets such as more or better livestock, more land, seeds and fertilisers. Quite apart from carrying repairs and running costs and charges for depreciation and interest, the purchaser of machinery or buildings may well be sacrificing immediate income together with his prospects of growth of capital and increased future income.

Purchase of more machinery or better buildings by a well-established farmer may be perfectly justifiable on the grounds of making life easier when the expenditure can be charged against heavy taxes paid on large profits; it is quite another matter for the young man who is just beginning and who needs every penny he can raise for the purchase of livestock and to produce the wherewithal to feed them. Frequently the best course in the latter case is to cut capital outlay on machinery to the bare minimum by employing a contractor and to invest the money more profitably elsewhere.

The simple truth is that the more a crop is handled, the greater are the labour and machinery requirements and the higher are the resultant costs. The easier and cheaper alternative to the use of hand or machine labour is therefore to eliminate the need for as many tasks as possible; examples of this approach are the techniques of direct drilling and grazing or utilisation of the crop in situ.

The numerous methods of production are summarised diagrammatically in Fig 7. The large number of possible combinations indicate the complexity of the problem. Only when the precise combination of methods is known is it possible to arrive at sensible estimates of production costs and capital requirements. Once again the danger of 'transplanting' generalised costs becomes only too evident.

The approach to production costs and investment depends to a large extent on whether the crop is to be harvested and stored or utilised in situ. Harvesting and storage usually requires a much heavier investment in machinery and equipment and tends to result in a 'high cost–high output' system; the aim must then be to obtain a really heavy yield of digestible fodder and to utilise it efficiently, which is the 'maincrop' approach.

In contrast, grazing or utilisation in situ is a 'low cost or minimal expenditure' system and is typical of the approach when growing catch crops. Really high yields are not necessary as waste tends to

FIGURE 7. Diagrammatic Summary of Methods of (A) Production and (B) Utilisation of Fodder C

(A) METHODS OF PRODUCTION

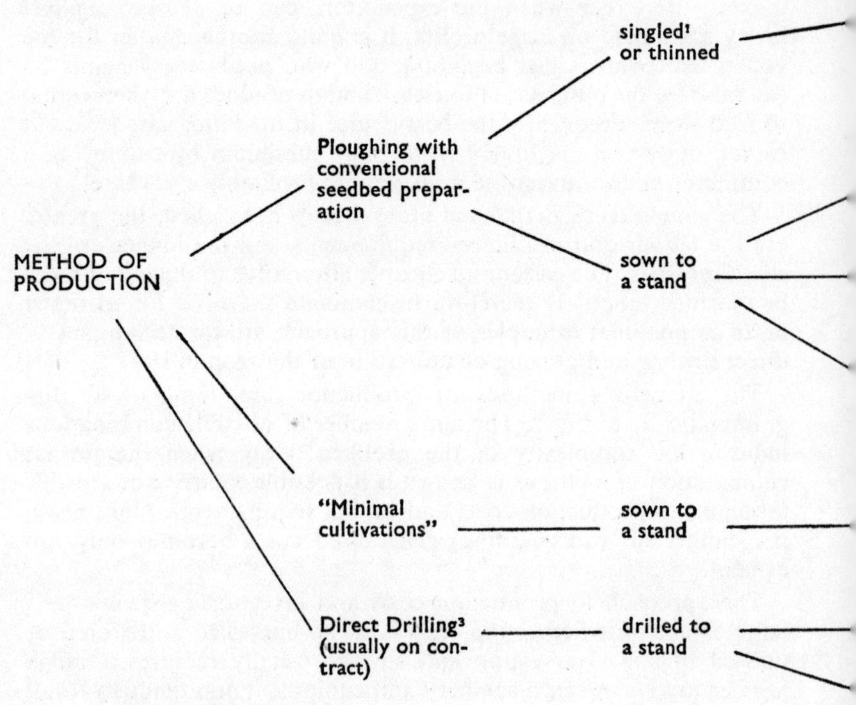

NOTES—(1) Crop sprayer or band spraying equipment usually necessary.
      (2) Crop sprayer usually necessary.
      (3) Desiccation of sward necessary before drilling.

	Equipment required for sowing, singling or transplanting	*Suitable crops*
	Transplanted	Flatpoll cabbage
'e rows ridges 28"	Precision drill + hand singling or Precision drill + down the row thinner or Turnip barrow + hand singling	Turnips, swedes, mangolds and fodder beet
le¹ rows 32"	Precision drill— drilled to a stand	Maize
le¹ rows 32"	Semi-precision or root drills Corn drills [4,5]	Kale, maize if no precision drill available
rows broad- t	Corn drill,[5] broadcast or manure dis- tributors	Stubble turnips, rape, kale, rape kales, fodder radish, rye, Italian ryegrass
rows broad- t	Corn drill[5] Broadcast or manure dis- tributor	
14" rows⁴	Hired direct drill or disc corn drill in softer stubbles[5]	Swedes,[6] kale or swedes and kale
'" rows		Stubble turnips, rape, Hungry gap or rape kale, fodder radish, rye, Italian ryegrass

(4) Unwanted rows blocked off.
(5) Inert "filler" may be necessary for small seeds.
(6) Direct drilled swedes for folding only.

FIG. 7 *(continued)*

(B) METHODS OF UTILISATION

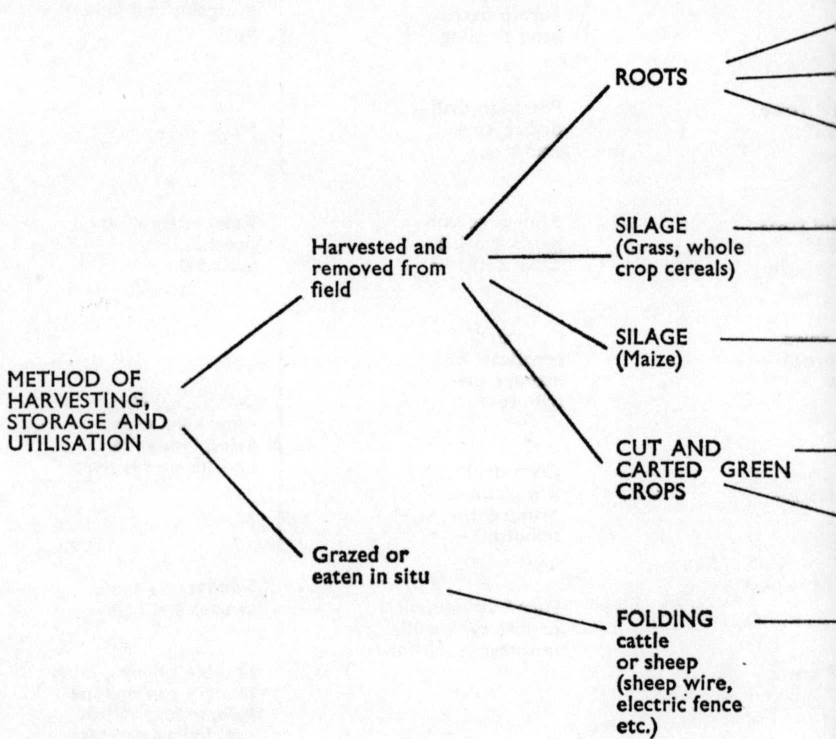

esting
nique
equipment

CLAMP,
STORE OR
BUILDING

pulped or pulped
and mixed with
other foods

carted out and fed
whole

removed mechanically
and "self feed"

and

per and tailer

plete root
ester

Turnips carted
and fed fresh

ple flail harvesting
em or full chop
esting system

cision chop smaller
ester or adapter
arge machine

SILO (Tower,
pit or
bunker)

self feed

easifeed

mechanical feeding
systems

O GRAZING—forage harvester—grass, kale,
green maize

HAND—cabbage and kale

des and turnips, rape, kale, rye and Italian ryegrass

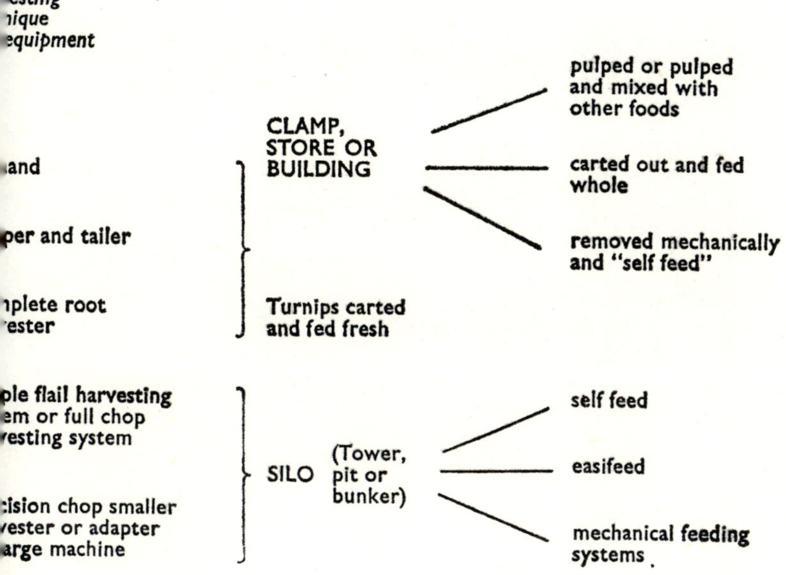

o

TABLE 32. Costs of Swedes Per Acre at Various Stages According to Crop Treatment

	Crops topped and tailed by hand Group 1 £	Mechanically Harvested Topped and tailed Group 2 £	Mechanically Harvested Complete Harvester (a) by Contractor Group 3 £	Mechanically Harvested Complete Harvester (b) by Farmer Group 4 £	Folded by Sheep Singled Group 5 £	Folded by Sheep Not Singled Group 6 £
Cost of crop in ground	18·76	20·80	16·44	14·51	20·17	13·33
Cost of singling and all row crop work	8·81	7·66	9·07	8·48	6·97	3·16
Cost of crop ready to harvest	27·57	28·46	25·51	22·99	27·14	16·49
Rent charge	5·00	5·00	5·00	5·00	5·00	5·00
Specialist machinery charges	1·56	3·77	1·52	3·67	1·03	0·75
Cost of harvesting and storage	18·40	11·86	15·96	6·68	—	—
Total cost/acre excluding overheads	52·53	49·09	47·99	38·34	33·17	22·24
Total cost/ton excluding overheads (30 ton yield)	1·75	1·64	1·60	1·28	—	—
Total cost/acre overheads @ £10	62·53	59·09	57·99	48·34	43·17	32·34
Total cost/ton overheads @ £10 (30 ton yield)	2·08	1·97	1·93	1·61	—	—
Total cost/acre overheads @ £20	72·53	69·09	67·99	58·34	—	—
Total cost/ton overheads @ £20 (30 ton yield)	2·42	2·30	2·27	1·94	—	—

Extracted from Reid, J. and Hunt, R. W. T., 'Turnips a Declining Crop?' West of Scotland Coll. Agric. Econ. Dept. Rep. No. 134. pp. 31.

e excessive when folding in wet weather. Extra machinery and
abour costs are kept to an absolute minimum by making maximal
use of machinery and facilities already on the farm and by techniques
such as direct drilling and drilling to a stand.

Broadcasting or drilling seed after a crop such as early potatoes,
when the soil is already loose and only requires levelling and
firming to produce a seedbed and undersowing Italian ryegrass in a
cereal are further examples of low cost techniques.

The much lower cost of kale and swedes utilised in situ is evident
from Tables 29 and 32. The precise problems vary according to the
type of crop.

Stored crops: Roots grown with conventional cultivations

In the past the 'problem' tasks with roots were singling, weed
control, harvesting and storage; labour requirements were high. The
total mechanisation of the crop is now possible and roots need no
longer be a 'labour intensive' crop but there are snags. The large
reduction in labour requirements as the operations used in swede
production are progressively eliminated are shown in Table 33.

**TABLE 33. Labour Requirements for Drilling, Singling and Harvesting
Swedes and Turnips**

DRILLING AND SINGLING

Operation	Output acres/hour	Labour requirement (man hours/acre)
Precision seeder	0·75	1·33
Down the row thinner[1]	0·80	2·50 (5·00)[2]
Hand hoeing after precision seeder	—	17·83
Inter-row work after precision seeder	1·00	2·00 (4·00)[2]

1. Two men used for this operation.
2. Bracketed figures relate to operations carried out twice.
The above is extracted from Pascal, J. A. The Mechanisation of Growing and
Harvesting Brassica Fodder Crops". Paper No. 3. The future of Brassica Fodder
Crops. Occ. Publ. Rowett Res. Inst. No. 2.

*Approximate labour required (man hours per acre) for sowing, singling
and inter-row work in swedes based on the above information*

Precision drilling, inter-row cultivations (twice) and singling by hand	23·0
Precision drilling, spraying and singling by hand	19·5
Precision drilling, inter-row cultivations (twice) and down the row thinning (twice)	10·5
Precision drilling, down the row thinning (twice) + overall spraying*	7·0
Precision drilling to a stand + overall spraying*	2·0

* Assuming 1 acre is sprayed in 0·4 hours.

HARVESTING†

Lifting, topping, tailing and carting:	Total labour requirement per acre	
	man hours	tractor hours
All hand work	40·0	12·9
Topping and tailing machine	20·3	14·6
Complete harvester	16·3	13·0

† Extracted from Reid, J. and Hunt, R. W. T. 'Turnips a Declining Crop?'
Rep. No. 134. Econ. Dept. W. Scot. Agric. Coll. pp. 31.

Sometimes the difficulty is to combine a low labour input with good weed control, a high plant population and maximum yield. Precision drilling followed by chopping out unwanted plants is reliable but requires a considerable amount of labour; this method is still widely used in Scotland. The use of down-the-row thinners reduces the labour requirement considerably; the plants may then be spaced somewhat irregularly but yield is unaffected by irregularity provided that an adequate plant population is obtained. The functioning of the thinner is frequently improved by the use of a herbicide.

Drilling to a stand with a precision drill and an appropriate herbicide is obviously by far the most attractive method in terms of labour economy. The frequently high cost of the herbicide must be set against the cost of the labour saved by this technique. The system has worked well in many instances, but the difficulties with weed control, discussed in Chapter 9, poor seedbeds or loss of plant from pigeon attacks have presented problems on some farms and in some areas. The choice of row width is limited by the minimal width required by the root harvester, which varies according to the make of the machine. Direct drilling is not suitable for crops to be harvested mechanically.

The mechanisation of the swede harvest results in a heavy reduction in the manual labour requirement but not in the number of tractor hours. The usual problem is to justify the cost of the machine. When substantial acreages of crops such as sugar beet are grown there is no difficulty but when fodder roots are the sole root crop there may not be a large enough acreage of roots to justify the purchase of a complete harvester.

When swedes only are grown the results of a recent investigation indicate that four acres or more can justify the purchase of a topping and tailing machine (stated price £150) in preference to hand lifting and topping; 15 acres or more can justify a complete harvester (stated price £400—current cost circa £450) in preference to hand lifting and topping and 60 acres or more justify the purchase of a complete harvester in preference to a topping and tailing machine. The position may be altered by a variation in the relative costs of manual labour and the price of a harvester.

Table 32 indicates that it is somewhat cheaper to employ a contractor with a complete harvester than to use manual labour, even assuming the latter can be obtained. The financial advantage of the harvester really becomes apparent when the machine is farmer-owned and syndicates are the obvious answer where the acreage is insufficient to justify a purchase by a single grower; assuming a syndicate of four growers purchases a complete harvester at £450, the individual capital outlay will be £112.50, which is less than would be spent if each man bought his own topping and tailing machine.

Cost is not the only factor that will influence the decision to purchase a harvester. If a contractor is to be employed it is essential that there is a reliable man in the district and that he will come when he is wanted. The timing of root harvesting is nowhere as critical as with hay, grass silage or corn harvest and the time at which a contractor arrives is of less importance. At the same time, the relatively long period over which the work may be spread is very favourable to ownership of a complete harvester by a syndicate.

When a complete harvester is brought in it is necessary to cater for the increased rate of working. Adequate labour, tractors and trailers should therefore be provided to handle the high rate of output and to ensure that there are no 'bottlenecks'.

The saving in the cost of casual labour as a result of mechanisation can be measured readily in terms of cash. The benefits are not always so obvious where regular labour is concerned and frequently do not result in any reduction in total staff. The investment must then be justified by speeding up the work and releasing labour for other more productive jobs. The resultant improvement in timeliness in carrying out livestock tasks and field operations increases farm output and frequently makes it possible to sow earlier and to increase the yield and the acreage of the more profitable crops such as winter wheat. Labour is worth much more than its nominal cost at peak periods.

The elimination of drudgery becomes of increasing importance as highly skilled and experienced workers become more difficult to obtain. Apart from the fact that the men can be used far more profitably on other jobs, the cold wet misery of hand work in a root field on a wet or frosty morning increases the incidence of sickness and is hardly conducive to retaining a highly skilled staff.

The high productivity and excellent digestibility of mangolds and fodder beet is beyond doubt. The main question is whether these crops can be handled economically with high labour and machinery costs. There is no problem with mangolds or fodder beet when other crops requiring a precision drill and root harvester are already on the farm. Frequently, however, these crops are grown in small blocks of an acre or so and are the only roots grown; such small acreages cannot justify mechanisation and unless adequate contract services or manual labour are available, these crops must, of necessity, disappear.

Stored crops: silage

When cut for tower silage, maize and whole crop cereals are

easier to handle than grass in a number of respects. The whole yield of the cereals is taken in a single cut and as wilting is not required direct cutting is possible; the chopped cereal material is also less sticky and less inclined to cause mechanical blockages. While there is a useful saving in labour, the main advantage in the handling of these crops is that they are harvested at a different time of the year from the main silage making period in May.

Most farmers cut their grass for silage with a simple flail type forage harvester. This type of machine can be used for cereal silage made in pits or bunkers but is of course quite unsuitable for maize, for which the following harvesting machinery is available:

(1) Self propelled forage harvesters which are only suitable for contractors and very large scale production.

(2) A pulled precision chop forage harvester with one or two row maize attachments.

(3) A single row tractor mounted maize chopper. The most suitable choice depends on the circumstances.

When a precision chop forage harvester is already in use or can be used for grass, the cheapest course would probably be to purchase a single row attachment; twin row attachments are more expensive and only appropriate to large scale production.

The special purchase of a precision chop forage harvester with adaptations can only be justified for really large areas of maize or by a contractor. Cost, including maize adaptors, varies from approximately £1,000–£2,350 and £1,500–£3,000 for harvesters with one and two row capacities respectively. Single-row maize choppers are most appropriate to normal farm conditions; cost varies from approximately £550–£850. These machines are either side mounted or mounted on the three-point linkage and highly manoeuvrable, especially under difficult conditions.

It has been suggested that around 50 acres of maize are needed to justify a purchase. Assuming depreciation over five years and an interest rate of 8 per cent the annual charge for depreciation and interest alone is £180. With a contract price of £4 per acre some 45 acres of maize would be needed to break even, without making any allowance for repairs and running costs. The relatively long period over which silage maize may be harvested is particularly favourable to syndicate ownership or employing a contractor.

The rate of output of any given machine depends on the soil conditions, the density of the crop and the power of the tractor. It is essential to have adequate power and to provide enough trailers to clear the harvester; the trailers should be fitted with tops to avoid wastage. Single row maize choppers will generally cope with 20–30 tons of green material per hour and forage harvesters with a single row adaptor about 5 tons more. In North Germany twin row pulled forage harvesters will deal with 60 tons of green material per hour. Silage maize harvesters are also suitable for handling kale grown in rows for 'zero grazing'.

Storage and handling

Economy and capital cost of handling and storage facilities vary enormously according to the type of crop and the degree of sophistication of the facilities provided. Complicated arrangements add greatly to the cost of the fodder and all proposals for new plant and buildings should be costed very carefully before additional expenditure is incurred. Wherever possible simple and versatile facilities should be provided. The root crops are particularly favourable in this respect as they require only the simplest storage arragements.

Winter grazing or utilisation of green crops in situ can be extremely awkward where very large herds of cows are concerned. When silage making machinery is already on the farm, 'zero grazing' or cutting and carting involves little or no additional capital cost; kale may well be worth considering in these circumstances.

Crops utilised in situ

Utilisation in situ allows full scope for cutting labour costs to the absolute minimum; drilling to a stand and direct drilling are well suited to this type of production. Apart from the contract price of the drilling which usually varies from £3 to £4 per acre according to the acreage drilled and whether it is grass or stubble, the only farm labour necessary with direct drilling is that required for spraying, distributing fertiliser and rolling once after the drill; a total of $1–1\frac{1}{2}$ hours per acre should generally suffice, according to the size of field and distance from the buildings. When swedes are direct drilled their labour requirement is little different from that of rape or kale but their potential output is usually higher.

The costs of production of direct drilling, which is largely represented by the variable costs compare very favourably with those of other methods of production.

Choice of fodder

Cost is only one of the factors to be taken into account when selecting the source of fodder; the crops chosen must be reliable in the circumstances of the particular farm and they must fit smoothly into the farming system. Soil, climate, pests and diseases usually limit the choice. Wet and very heavy soils are best left down to grass; a large number of farms must necessarily remain all grass farms.

The alternative sources of energy to fodder crops are grass and cereals; the advantages and limitations of grass have already been discussed. The extent to which cereals may be used depends on the price. If high cereal prices prevail, cheaply produced fodder crops become a very attractive substitute for at least part of the cereals in the ration. The opportunity for using fodder crops depends on the class of livestock and the system under which they are kept.

The output of high yielding autumn calving dairy cows necessitates the use of a 'concentrated' ration; the intake of 'bulk' must therefore be restricted. High quality hay and silage can reduce the concentrate requirement but cannot eliminate it; in the early stages of the lactation quality of food rather than strict economy of cost is the factor of prime importance. Maize and whole crop cereal silage vary less in quality than grass silages and are therefore potential replacements or supplements for all but top quality grass silage. Ease and economy of handling are of considerable importance with in-wintered animals.

Intensively kept stock with a high cash output can pay for an appropriate investment in buildings and machinery. However, in spite of the limitations imposed on their use, green fodder crops can be particularly useful for cutting down on the need for conservation and shortening the length of the winter feeding period. Kale grazed judiciously in the autumn and rye and Italian ryegrass for early bite can be invaluable, especially if the quality of the hay is not as good as it might be.

The cost of milk production is all-important with the spring calving herd, so that concentrate usage must be limited substantially to steaming up and to the first part of the lactation before spring grass is available. Rye and Italian ryegrass can provide useful grazing on well-drained soils before any other grass comes; if poaching occurs with these crops they can be ploughed readily and their use safeguards the longer leys and permanent grass.

Early sown kale is invaluable for feeding in a dry August and can be followed by a succession of green fodder crops to supplement the grass. It is best to fold grass and fodder crop concurrently, giving the fodder crop in the morning and the grass overnight. The supply of grass in the back end can be extended by using Italian ryegrass.

Yarded fattening bullocks are entirely dependent on stored fodder, which must be of high quality and digestibility. All too often the quality of hay or grass silage is inadequate for the job and high cereal prices make heavy supplementation with concentrates an uneconomic proposition. Provided that appropriate handling facilities are employed swedes or maize silage, which are both highly palatable, digestible and of uniform quality, form an admirable basis for the fattening ration.

The problem is somewhat different with stock such as sheep and single suckling cows, which frequently give a relatively low output per acre. These stock will not usually stand heavy investment in housing or in machinery and buildings for conservation and storage. Facilities must be cheap and simple and the fodder should be utilised in situ wherever possible.

Roots and green fodder crops are particularly applicable to sheep, as they are highly digestible and can be grown and utilised cheaply. The most suitable crops depend on the system of flock management and the time of year the fodder is required.

Flocks lambed between October and Christmas for early fat lamb production (Dorset Horn and 'Down' type flocks) are usually heavily dependent on roots and green forage crops. Rape, turnips, kale and perhaps ryegrass are appropriate for use before Christmas but thereafter frost hardy crops such as swedes and thousand head kale are required. Rye or Italian ryegrass make excellent 'follow on' crops in the spring where required. This is a 'high cost—high return' system and it is essential to ensure that the lambs are ready before the high prices decline. As speed of growth of the lambs is all-important, concentrates are frequently creep-fed to the lambs.

The alternative is the 'low cost' system, in which lambing occurs in late February, March or even in early April for store lamb production. After lambing, the ewes and lambs go straight out to grass, where they are kept intensively until weaning. The lambs are then moved on to fresh young aftermath and the remaining lambs are finished on rape or rape and turnips in the autumn.

Although this is a 'cheap' system, the standard of livestock husbandry must be high. On no account must the ewes be under-nourished before lambing or pregnancy toxaemia ('twin-lamb' disease) is likely.

Swedes supplemented by autumn grown grass or hay form a useful pre-lambing diet in January, February and early March. A small quantity of concentrates may also be given to the ewes before and immediately after lambing. The amount of concentrates given depends on the proportion of twins expected, the quality of the feed and the severity of the climate; apart from this limited usage, concentrate feeding is generally uneconomic. The provision of rye or Italian ryegrass ensures that there is an early supply of fodder to promote the ewes' milk supply and get the lambs away to a really good start. If this is not available clamped roots or maize silage is valuable.

If nothing worse occurs, the feeding of poor hay before and during the lambing period results in ewes without any milk. The lambs must also receive proper treatment against worms during the summer and have been immunised against pulpy kidney disease before they go on to the rape in the autumn.

Root and green forage crops can play a useful part on stock rearing farms. Provided they are not overcropped, swedes can be used to increase the productivity of suitable 'in bye' land and cut down on the overwintering costs of cattle and sheep. Stubble catch crops have proved invaluable for autumn-calved single suckling cows in the south of England. The cows milk well and, by the spring, the calves are forward and well grown and well able to make full use of the grass when it comes.

The problems of the intensive cereal grower are many. Leaf borne diseases, grassy weeds and difficulties with soil structure are taking their toll and in many cases yields are static or even declining while costs such as labour, fertilisers, machinery and repairs are rising sharply, so that margins are being squeezed hard in many cases. Alternative crops for direct sale are difficult to find and all too frequently are over-produced. Cereal prices may well rise considerably but whether a rise in price alone is enough to sustain farming systems which are of dubious technical and economic viability is quite another matter.

The introduction of intensively kept livestock can doubly benefit the cereal grower. Quite apart from the increase in cash output from the livestock, the plant and animal residues left from folded roots or green crops benefit the following corn crop considerably and reduce the amount of artificial that it is necessary to apply. Weed control is also made easier. Provided always that the stubble is free from couch before a catch crop is planted, the canopy of a leafy brassica crop shades the ground and prevents the further growth of grassy weeds in the autumn. When grown as a 'break' crop fodder crops can also be valuable for the control of wild oats.

Economic situations change and with them whole farming systems. The old arable systems of folding flocks of sheep on hand singled roots and the yard fattening of beef foundered on heavy labour demands and low meat and cereal prices. The high labour requirements for fodder crops no longer apply.

With a rising population and heavy home and EEC export demands, there is ample room for a large increase in animal production on mixed farms, uplands and predominantly arable farms alike.

The knowledge that livestock and cereals fit together admirably is not new. The chalks, the limestones and the lighter loams are freely drained and well suited to cattle and sheep alike; these soils have been recognised as ideal 'sheep and barley land' for a very long time.

There is considerable unused productive capacity in Britain. We have the knowledge, we have the resources and we have the opportunity; there is no need to wait.

APPENDICES

APPENDIX 1

Summary of Cropping Situations and Sowing Dates at Which Root and Brassica Green Forage Crops May Be Used to Provide Fodder at Different Times of the Year

G = General recommendation
N = N. England, Scotland and exposed districts

S = S. of England and mild or sheltered districts
( ) = Denotes limited application or reduced reliability

* = Crop fed from store
† = Crop frequently fed from store, especially in cold districts

EARLY SOWN/PLANTED MAINCROPS
Sown: April and May

Latest possible sowing/planting date			Month(s) during which crop is suitable for use										Remarks
N	S	Crop	Aug	Sep	Oct	Nov	Dec	Jan	Feb	Mar	Apr	May	
As early as possible	20 May	Mangolds and fodder beet*						G	G	G	G	G	a
31 May	—	Swedes†			N	N	N	N	N	N	(N)		b
31 May	—	Turnips, hardy yellow†			N	N	N	N	N				c
7 May	31 May	Cabbage, flat poll (aut. sown)			G	G	S						d
31 May	—	Kale, marrow stem or Maris Kestrel	G	G	N	N							e

Remarks:
a   Unsafe to feed before December. Almost invariably essential to store. Rarely grown In Scotland and not really suitable North of Darlington, Yorks
b   A crop of swedes is frequently fed partially from the field and partially stored, the proportions depending on variety and district. Crops for later feeding may be wholly stored; the last date for use or storage depends on the hardiness of the variety and the district. Only the latest varieties are reliable for use in April and then in colder areas.
c   Largely grown as a maincrop on poorer soils in Scotland; storage as for swedes. Little grown elsewhere.
d   Can be planted in early June in South but only heavy crops can be justified. Not very frost resistant; may stand into new year in milder districts.
e   Only a limited acreage should be sown early in the South of England and then suited mainly to drier situations where the grass is inclined to burn in dry weather.

APPENDIX 1 (contd)

Summary of Cropping Situations and Sowing Dates at Which Root and Brassica Green Forage Crops May Be Used to Provide Fodder at Different Times of the Year

G = General recommendation
N = N. England, Scotland and exposed districts

S = S. of England and mild or sheltered districts
( ) = Denotes limited application or reduced reliability

* = Crop fed from store
† = Crop frequently fed from store, especially in cold districts

LATE SOWN MAINCROPS, EARLY PLANTED CATCH CROPS FOLLOWING EARLY POTATOES, PEAS, LEY, ETC.
Sown: June–Mid-July

Latest possible sowing/ planting date N	S	Crop	Aug	Sep	Oct	Nov	Dec	Jan	Feb	Mar	Apr	May	Remarks
—	30 June	Swedes						S	S	S			f
—	30 June	Turnips, hardy yellow					S	S	S				
15 June	15 July	soft yellow		N	N	G	S						g
15 June	15 July	white (bulbs)	N	N	G	S	S						
Unsuitable	15 July	Kale, Marrow Stem			S	S	S						h
10 June	15 July	Maris Kestrel			G	G	G	S	S	S			h
10 June	15 July	Thousand Head			Maris Kestrel usually pre-ferable			S	S	S			i
15 July	15 Aug.	Forage rape—sizeable crop		G	G	G							j

f South of England only, where early sowings are especially susceptible to mildew.
g Grown for full crop of bulbs. Late sowings permissible up to mid-July in north and early August in south but yield will be reduced correspondingly. Plant density should also be increased with delayed sowing. Mixture of Italian Ryegrass and hardy yellow turnips may be sown June to early July according to district.
h Main sowings in South of England. Maris Kestrel to be preferred to Marrow Stem kale.
i Thousand Head kale is hardy but can be heavily damaged in exposed situations and severe winters. Stored roots are usually much safer in exposed northern areas. Should replace Marrow Stem kale for earlier use in exposed hilly districts. Little grown in Northern England.
j Earlier sowings of rape in this period more suited to hill areas in Wales and in Northern England; swedes or kale give a better yield when sown at at the same date in the South of England.

APPENDIX I (contd)

Summary of Cropping Situations and Sowing Dates at Which Root and Brassica Green Forage Crops May Be Used to Provide Fodder at Different Times of the Year

G = General recommendation
N = N. England, Scotland and exposed districts
S = S. of England and mild or sheltered districts
( ) = Denotes limited application or reduced reliability
* = Crop fed from store
† = Crop frequently fed from store, especially in cold districts

LATE SOWN CATCH CROPS FOLLOWING CEREALS OR LATE BROKEN GRASS
N.B. ONLY after winter cereal in NORTHERN areas and late districts
Sown: Mid-July–early September

Latest possible sowing/planting date N	S	Crop	Month(s) during which crop is suitable for use										Remarks
			Aug	Sep	Oct	Nov	Dec	Jan	Feb	Mar	Apr	May	
—	15 July	Turnip, hardy and frost resistant yellows					S	S	S	(S)			k
—	30 July	white/continental stubble—some bulb production			G	G	S						l
7 Aug.	10 Sep.	white/continental stubble—mainly leaf production			G	G	S						m
—	30 Aug.	Hardy Green Round						S	S	S			n
—	30 July	Kale, Marrow Stem				Forage rape usually preferable							
—	30 July	Maris Kestrel						S	S	S			o
—	30 July	Thousand Head						S	S	S			
Unsuitable	15 Aug.	Rape Kale and Hungry Gap				Forage rape preferable		S	S	S	(S)		p
7 Aug.	10 Sep.	Forage rape—lighter crop			G	G	S	(S)	(S)	(S)			q
7 Aug.	30 Aug.	Fodder Radish	(G)	(S)									r

(Kale varieties Marrow Stem, Maris Kestrel, Thousand Head: Light 'sheep' crops only obtained)

k   Only in areas with open autumn. Crop tends to be lighter than with earlier sowings—may be included in mixtures. Hardier soft varieties, e.g. Tuckers Green Top Yellow best for these later sowings.
l   Last date applicable to bulb production.
m   Not winter hardy—use before Christmas.
n   Should always be used in preference to non hardy types for feeding after Christmas; grown mainly for top production, roots are small.
o   ONLY in areas with open autumn; sowing should be completed before the end of July. Mainly for provision of hardy sheep keep. Crop usually rather light.
p   Hungry gap may be used in autumn if required but rape gives rather better yield. Rape kale is a little hardier and only sown for spring use. Both susceptible to mildew and should not be sown before late June or early July. Outyielded by B. oleracea kales at earlier sowings.
q   Use after Christmas ONLY in mildest areas—even then considerable risk of severe frost damage.
r   Not frost hardy and does not appear to do well in many northern situations—use limited.

N.B. This table is essentially a generalised summary and there is likely to be considerable divergence in last sowing dates and period of utilisation appropriate to different

APPENDIX 2

VARIETIES OF ITALIAN RYEGRASS
(arranged in order of heading)

Varieties recommended by NIAB

Grasslands Manawa — Persistency good but frost hardiness poor. Only suited to mildest areas in south and west of England.

Rv.P.
Lema
Combita
Optima
(previously Sceempter) — These varieties exhibit a high degree of winter hardiness and good all year round production Combita is especially hardy and gives the highest early spring and spring yields.

Varieties recommended by NIAB but becoming outclassed

Danish varieties:

E.F. 486
Leda Daehnfeldt
Lomi Trifolium
Prima Roskilde
Vejrup M.B. — These varieties show good winter hardiness with good early spring and spring production but are short lived and give inferior production to the recommended list after midsummer. Suitable mainly for catch crops where hardiness and early bite are required.

Tetila tetraploid — Gives later spring growth than many other varieties. Hardiness good.

Ellesmere — Bred from Italian and perennial ryegrasses but retains the Italian characteristics. Good all round performance and hardiness.

Aberystwyth S.22 — Owing to its susceptibility to damage in severe winters and relatively later start in the spring this variety is quite unsuited to Scotland and northern England but does well in south west England. S.22 flowers 3–6 days later than the other varieties.

Other varieties which may be considered if recommended varieties are not available: Hesa, Grasslands Paroa, Motterwitzer, Stormont Ibex.

APPENDIX 3

Table for Converting British and Metric Measurements

BRITISH TO METRIC

METRIC TO BRITISH

LENGTH

1 inch (in) = 2·54 cm	1 millimetre (mm) = 0·0394 in
or 25·4 mm	1 centimetre (cm) = 0·394 in
1 foot (ft) = 0·30 m	1 metre (m) = 1·09 yd
1 yard (yd) = 0·91 m	1 kilometre (km) = 0·621 miles
1 mile = 1·61 km	

Conversion Factors

inches to cm × 2·54	centimetres to in × 0·394
or mm × 25·4	millimetres to in × 0·0394
feet to m × 0·305	metres to ft × 3·28
yards to m × 0·914	metres to yd × 1·09
miles to km × 1·61	kilometres to miles × 0·621

AREA

1 sq. inch (in^2) = 6·45 cm^2	1 sq. centimetre (cm^2) = 0·16 in^2
1 sq. foot (ft^2) = 0·093 m^2	1 sq. metre (m^2) = 1·20 yd^2
1 sq. yard (yd^2) = 0·836 m^2	1 sq. metre (m^2) = 10·8 ft^2
1 acre (ac) = 4047 m^2	1 hectare (ha) = 2·47 ac
or 0·405 ha	

Conversion Factors

sq. feet to m^2 × 0·093	sq. metres to ft^2 × 10·8
sq. yards to m^2 × 0·836	sq. metres to yd^2 × 1·20
acres to ha × 0·405	hectares to ac × 2·47

VOLUME (LIQUID)

1 fluid ounce (1 fl oz)	100 millilitres (ml or cc) = 0·176 pints
(0·05 pint) = 28·4 ml	1 litre = 1·76 pints
1 pint = 0·568 litres	1 kilolitre (1000 litres) = 220 gal
1 gallon (gal) = 4·55 litres	

Conversion Factors

pints to litres × 0·568	litres to pints × 1·76
gallons to litres × 4·55	litres to gallons × 0·220

WEIGHT

1 ounce (oz) = 28·3 g	1 gramme (g) = 0·053 oz
1 pound (lb) = 454 g	100 grammes = 3·53 oz
or 0·454 kg	1 kilogramme (kg) = 2·20 lb
1 hundredweight (cwt) = 50·8 kg	1 tonne (t) = 2204 lb
1 ton = 1016 kg	or 0·984 ton
or 1·016 t	

Conversion Factors

ounces to g × 28·3	grammes to oz × 0·0353
pounds to g × 454	grammes to lb × 0·00220
pounds to kg × 0·454	kilogrammes to lb × 2·20
hundredweights to kg × 50·8	kilogrammes to cwt × 0·020
hundredweights to t × 0·0508	tonnes to tons × 0·984
tons to kg × 1016·0	
tons to t × 1·016	

USAGE OR YIELD PER ACRE AND PER HECTARE

BRITISH TO METRIC

1 gallon per acre = 11·2 litres per ha

1 pound per acre = 1·12 kg per ha

1 hundredweight per acre
    = 125 kg per ha
1 ton per acre   = 2500 kg per ha
    or 2·5 t per ha

METRIC TO BRITISH

1 litre per hectare   = 0·09 gal per ac
    or 0·7 pint per ac
100 litres per hectare = 9 gal per ac
1 kilogramme per
    hectare   = 0·9 lb per ac
    or 14·2 oz per ac
100 kilogrammes per hectare (1 quintal)
    = 0·8 cwt per ac
1 tonne per hectare = 8 cwt per ac
    or 0·4 tons per ac

Fertiliser Units

1 kg fertiliser unit (French unit) per hectare = 0·8 British fertiliser units per acre

Plant Population

No. plants per sq. foot to
No. plants per m² × 10·8
No. plants per sq. yard to
No. plants per m² × 1·20
No. plants per acre to
No. plants per ha × 2·47

No. plants per sq. metre to
No. plants per ft² × 0·093
No. plants per sq. metre to
No. plants per yd² × 0·836
No. plants per hectare to
No. plants per ac × 0·405

APPENDIX 4

ADDITIONAL READING
Books, review articles, etc., arranged according to chapter

Chapter 1
Anon. (1968). 'A Century of Agricultural Statistics—Great Britain 1866–1966'. Min. Agric. Fish & Fd. & Dept. Agric. & Fish. Scotland. H.M.S.O. London. pp. 129.

Chapter 2
Greenhalgh, J. F. D. and Hamilton, M. (1971). 'The Future of Brassica Fodder Crops'. Occ. Publ. Rowett Res. Inst. No. 2. pp. 67.
  Proceedings of a discussion meeting held at the Rowett Research Institute, Aberdeen on 11 June 1971.
  (Nine papers covering all aspects of brassica root and fodder crops; bibl. 111).
Holliday, R. (1960). 'Plant Population and Crop Yield Parts 1 and 2'. Field Crop Abstr. **13** No. 3. 159–67 and No. 4 247–54 (Review article bibl. 80).
Nat. Inst. Agric. Bot. (1971). 'Varieties of Green Fodder Crops'. Farmers leaflet No. 2. pp. 15.
Willey, L. A. (1964). 'The Kale Crop in Great Britain'. Fld. Crop Abstr. **17**. No. 1. 1–5 (Review article bibl. 64).

Chapter 3
Harvey, P. N. (1969). 'Spring Mechanisation of Sugar Beet'. NAAS Quart. Rev. No. 83. 105–115 (Review article bibl. 6).
Hull, R. and Jaggard, K. W. (1971). 'Recent Developments in the Establishment of Sugar Beet Stands'. Fld. Crop Abstr. **24**. No. 3. 381–88 (Review article bibl. 107).
McLean, K. A. (1970). 'Selective Thinning of Row Crops'. NAAS Quart. Rev. No. 89. 23–32 (Review article bibl. 5).
McNaughton, I. H. and Thow, R. F. (1972). 'Swedes and Turnips'. Fld. Crop Abstr. **25**. No. 1. 1–12 (Review article bibl. 181).
Morrison D. et al. (1969). 'Turnips and Swedes'. Advisory Bulln. No. 6. Dept. Agric. Fish. Scotland. HMSO Edinburgh. pp. 33 (bibl. 14).
Nat. Inst. Agric. Bot. (1972). 'Varieties of Fodder Root Crops'. Farmers leaflet No. 6. pp. 9.

Chapter 4
Alderman, G. (1963). 'Mineral Nutrition and Reproduction in Cattle'. Vet. Record **75**. 1015–80. (Bibl. 21).
Blood, D. C. and Henderson, J. A. (1968). 'Veterinary Medicine'. 3rd Ed. Baillère Tindall and Cassel, London. pp. 927.
Dawson, F. L. M. (1970). 'Infertility in Beef Cattle'. Vet. Bulletin **40**. No. 11. Comm. Bureau An. Health, Weybridge, Surrey. (Review Article Bibl. 136).
Fairbairn, C. B., Smith, A. E., and Hart, R. (1971). 'Brassica Waste as a Feed for Beef Animals and Dairy Cows'. ADAS Quart. Rev. No. 3. Winter 1971. 134–40 (Review article bibl. 15).
Forsyth, A. A. (1968). 'British Poisonous Plants'. Bull. No. 161. Min. Agric. Fish & Fd. HMSO London. pp. 131.
Greenhalgh, J. F. D. (1971). 'Problems of Animal Disease'. Occ. Publ. Rowett Res. Inst. No. 2. 56–64 (bibl. 28).
Jones, F. G. W. and Jones, M. (1964). 'Pests of Field Crops'. Edward Arnold (Publishers) Ltd. London. pp. 386.
Jones, F. G. W. and Dunning, R. A. (1969). 'Sugar Beet Pests'. Bull. No. 162. Min. Agric. Fish & Food. HMSO London. pp. 108.
Martin, H. (Ed.) (1972). 'Insecticide and Fungicide Handbook for Crop Protection'. Blackwell Scientific Publications, London. pp. 416.
Rhodes, J. (1972). 'Swedes in the Seventies—A Fully Mechanised Break Crop'. Agriculture. London. 189–192.

Underwood, E. J. (1966). 'The Mineral Nutrition of Livestock'. FAO Comm.
Agric. Bureaux. pp. 237 (Bibl. 380).
Wilson, J. G. (1966). 'Bovine Functional Infertility in Devon and Cornwall.
Response to Managanese Therapy'. Vet. Record. **79**. No. 20. 562–66 (Bibl. 10).
Chapter 5
Charles, A. H. (1958). 'The Effect of Undersowing on the Cereal Cover Crop'.
Fld. Crop Abstr. **11**. No. 4. 233–9 (Review Article Bibl. 47).
Francis, A. L. and Dover, P. A. (1965). 'Undersowing Small Seeds Under a
Cereal Cover Crop—The Effect on Yield of Cereals and the Establishment of
the Ley'. Experimental Husb. HMSO London. No. 12. 89–104 (bibl. 7).
Holliday, R. (1956). 'Fodder Production from Winter Sown Cereals and its
Effect on Grain Yield. Parts 1 and 2'. Fld. Crop Abstr. **9**. No. 3. 129–35 and
No. 4. 207–13 (Review article bibl. 57).
Nat. Inst. Agric. Bot. (1971). 'Recommended Varieties of Grasses'. Farmers
leaflet No. 16. pp. 21.
Chapter 6
Anon. (1970). 'Maize as a Forage Crop'. Occ. Publ. Maize Development Assn.
Tunbridge Wells, Kent. pp. 23.
Anon. (1972). 'Growers Guide to Forage Maize Harvesters'. Maize Bull. No.
40. 8–11.
Bunting, E. S. (1968). 'Maize in Europe'. Fld. Crop Abstr. **21**. No. 1. 1–9
(Review article bibl. 159).
Harris, P. B. (1971). 'Developments in Making Whole Cereal Silage'. Ann.
Rep. No. 12. Bridgets Exp. Husb. Farm, Martyr Worthy, Winchester, Hants.
(MAFF) 23–30.
Lang, R. W. and Holmes, J. C. (1969). 'Changes in Yield and Quality of Barley
Grain and Straw During Maturation'. Experimental Husb. HMSO London.
No. 18. 1–7 (bibl. 19).
Nat. Inst. Agric. Bot. (1972). 'Recommended lists of maize varieties 1972'.
Supplement to Farmers leaflet No. 2—Varieties of Green Fodder Crops. pp. 4.
Newman, G. and Rieman, U. (1972). 'Silage Maize Harvesting'. Maize Bull.*
No. 40. 4–7.
Raymond, W. F. and Heard, A. J. (1968). 'The Ensilage of Whole Crop Cereals'.
Ceres. 1968. No. 4. 7–11 (Review article bibl. 7).
Sheldrick, R. D. 'Growing Silage Maize'. ADAS Quart. Rev. No. 4. Spring
1972. 177–186 (Review article bibl. 29).
Thomas, C. and Wilkinson, J. M. (1971). 'Maize Silage for Dairy Cows—Part 1.
The Effect of Silage Composition on Nutritive Value and on Milk Production.'
Maize Bull. No. 33. 5–6 and 12 (Review article bibl. 17).
Thomas, C. and Wilkinson, J. M. (1971). 'Maize Silage for Dairy Cows—Part 2.
The Effect of Supplementation of Maize Silage on Voluntary Intake and Milk
Production'. Maize Bull. No. 33. 5–6 and 12 (Review article bibl. 19).
Chapter 7
Agricultural Advisory Council. (1970). 'Modern Farming and The Soil'. HMSO
London. pp. 119.
Anon. (1970). 'Soil structure and Subsoiling'. Short Term Leaflet No. 114.
Min. Agric. Fish & Fd. London. pp. 10.
Anon. (1971). 'Getting down to Drainage'. Series of leaflets Nos. 1–11. Min.
Agric. Fish. & Fd. London.
Chapter 8
Toosey, R. D. (1971). 'Direct Drilling of Fodder Crops'. ADAS Quart. Rev.
No. 3. Winter 1971. 121–33 (Review article bibl. 36).
Chapter 9
Anon. (1972). 'List of Approved Products for Farmers and Growers'. Min.
Agric. Fish. & Fd. London.
Chancellor, R. J. (1959). 'Identification of Seedlings of Common Weeds'. Bull.
No. 179. Min. Agric. Fish. & Fd. HMSO London. pp. 72.
Chancellor, R. J. (1966). 'The Identification of Weed Seedlings of Farm and
Garden'. Blackwell Scientific Publns. Oxford. pp. 88.

Fryer, J. D. and Evans, S. A. (1968). 'Weed Control Handbook Vol. I—Principles. Fifth Edn.' Blackwell Scientific Publns. Oxford & Edinburgh. pp. 494.
Fryer, J. D. and Makepeace, R. J. (1970). 'Weed Control Handbook Vol. II—Recommendations. Sixth Edn.' Blackwell Scientific Publns. Oxford & Edinburgh. pp. 331.
Williams, J. H. (1970). 'Herbicides—Their Fate and Persistence in Soils'. NAAS Quart. Rev. No. 87. 119–132 (Review article bibl. 43).

Chapter 10
Anon. (1963). 'The Farm as a Business—1. Introduction to Management'. Min. Agric. Fish. & Fd. HMSO London. pp. 51.
Anon. (1965). 'The Farm as a Business—7. Aids to Management—Arable Crops and Grass'. Min. Agric. Fish & Fd. HMSO London. pp. 41.
Anon. (1967). 'Low Cost Production Report; LCP 67'. Milk Marketing Board, Thames Ditton, Surrey. pp. 60.
Anon. (1969). 'The Farm as a Business—6. Aids to Management—Labour and Machinery'. Min. Agric. Fish & Fd. HMSO London. pp. 51.
Anon. (1971). 'A Farm Business Management Handbook—First Edn.'. Agric. Econ. Divn., N. Scot. Coll. Agric. Aberdeen. pp. 207.
Anon. (1971). 'Results from Store Lamb Feeding and Finishing Enterprises 1969/70'. MLC Sheep Improvement Services, Meat & Livestock Comm. c/o An. Breed. Res. Org., Edinburgh, pp. 9.
Barnard, C. S., Halley, R. J. and Scott, A. H. (1970). 'Milk Production'. Iliffe Books Ltd. London. pp. 254.
Nix, J. S. (1971). 'Farm Management Pocketbook—Fourth Edn.' Sch. Rural Econ. & Related Studies, Wye Coll. Univ. London. pp. 135.
Reid, J. and Hunt, R. W. T. (1970). 'Turnips—A Declining Forage Crop?' Rep. No. 134 Econ. Dept. W. Scot. Agric. Coll. pp. 31.
Sturrock, F. (1971). 'Farm Accounting and Management—Sixth Edn.' Pitman Publishing. London. pp. 262.

GENERAL

Feeding
Evans, R. E. (1960). Rations for Livestock. Bull. 48. Min. Agric. Fish & Fd. HMSO London. pp. 134 (bibl. 16).
Russell, K. (1969). 'The Principles of Dairy Farming'. Farming Press Ltd. Ipswich. pp. 264.

Manuring
Anon. (1969). 'Lime and liming'. Bull No. 35. Min. Agric. Fish & Fd. HMSO London.
Cooke, G. W. (1964). 'Fertilisers and Profitable Farming—Second Edn.'. Crosby Lockwood, London. pp. 143 (bibl. 66).
Cooke, G. W. (1967). 'The Control of Soil Fertility—First Edn.'. Crosby Lockwood, London. pp. 526 (bibl. 661).
Wallace, T. (1961). 'The Diagnosis of Mineral Deficiencies in Plants by Visual Symptoms—A Colour Atlas and Guide—Third Edn.'. HMSO London.

---

*Bulletin of Maize Development Association, 16 Lonsdale Gardens, Tunbridge Wells, Kent.

INDEX